Praise for *Evolutionary Psychology: A Beginner's Guide*

"The *evolutionary* approach to the human mind is becoming more *and* more influential, but its claims and assumptions are often misun*derstood* by proponents and critics alike. This slim but impor*tant* volume gives us a very readable – and much needed – overvi*ew* of what evolutionary psychology is and is not."

Dr Daniel Nettle, Lecturer in Psychology, University of Newcastle upon Tyne, England

"I like *this* book very much. It is well-written and easy to read, and clears *up* many of the most persistent misunderstandings about evolu*tio*nary psychology."

Dr Susan Blackmore, author of *The Meme Machine*

... *psychology's key
... on human actions
... ural evolution
... n traits, our moral

... *e College, Michigan

RELATED TITLES FROM ONEWORLD

Artificial Intelligence: A Beginner's Guide, Blay Whitby, ISBN 1–85168–322–4
Genetics: A Beginner's Guide, B. Guttman, A. Griffiths, D. Suzuki and T. Cullis, ISBN 1–85168–304–6
Moral Relativism: A Short Introduction, Neil Levy, ISBN 1–85168–305–4
What Makes us Moral?, Neil Levy, ISBN 1–85168–341–0

evolutionary psychology

a beginner's guide

human behaviour, evolution and the mind

robin dunbar
louise barrett
john lycett

ONEWORLD
OXFORD

evolutionary psychology: a beginner's guide

Oneworld Publications
(Sales and Editorial)
185 Banbury Road
Oxford OX2 7AR
England
www.oneworld-publications.com

ISBN 1–85168–356–9

Typeset by Jayvee, Trivandrum, India
Cover design by the Bridgewater Book Company
Printed and bound in India by Thomson Press Ltd

contents

why do we need evolution?

When Charles Darwin launched his book *On the Origin of Species* on an unsuspecting Victorian audience in 1859, he triggered an intellectual shock wave that continues to send ripples around the world. One implication to which his book drew attention (though it was not one of Darwin's own insights) was the fact that we humans are very much members of the animal kingdom. We are members of the order Primates, the group that contains all the monkey and ape species. In the past decade or so, we have gone one step further than any of Darwin's contemporaries ever imagined in this respect. Genetic evidence has convincingly shown that, far from being a distant cousin of the monkeys and apes, we are very firmly embedded in the ape family as the sister species of the chimpanzees.

darwin and the mind

While Darwin's theory of natural selection has been truly revolutionary in the history of science, it was not Darwin's ideas on how species are formed that were to achieve such prominence in the long run. Looking back on the 150 years of intellectual history since *The Origin* was published, we can see a growing importance for his later books *The Descent of Man* (in which he explored sexual selection and reproductive behaviour) and *The Expression of the Emotions in Man and Animals* (in which he tackled the nascent field of psychology). Darwin's ventures into the field of behaviour were

1

much underrated – indeed, his theory of sexual selection, with its emphasis on the processes of mate choice, did not come to occupy the position of importance that it now does within evolutionary biology until more than a century after *The Descent of Man* was published. And in many ways, we are still absorbing the lessons of his work on emotions. But both books were extraordinarily prescient, in that Darwin put his finger on issues that have since come to be seen as fundamental to our understanding of human behaviour and the mind that underpins it.

The past three decades have witnessed an extraordinary explosion in our understanding of animal behaviour and its evolutionary components. This explosion has involved both the development of a very sophisticated body of theory, much of it underpinned by mathematical models and a volume of observational and experimental research on animal behaviour that would have excited the grand old man beyond measure. For it was Darwin's genius to bring together a powerful combination of acute theoretical insight with empirical tests using data from a wide range of species. Known as the comparative method, this has remained the methodological cornerstone of the evolutionary approach to this day.

While the study of non-human animals progressed apace from the 1970s, the extension of these ideas to human behaviour and psychology had to wait for the better part of another two decades before its own explosive take-off. In part this reflected a nervousness on the part of biologists towards dabbling in things human, but also the distrust in which social scientists had held evolutionary and biological ideas since the early 1900s. However, from the late 1980s onwards, evolutionary ideas began to be applied in earnest to the study of human behaviour and the human mind. This field is so new that its findings are only available in the more specialized journals. This book is an attempt to draw together some of the more salient findings from this research in a form accessible to the general reader.

Before we begin, we need to make clear what an evolutionary approach to human behaviour does – and does not – entail. The value of the evolutionary approach is that it provides us with a sound theoretical framework which enables us to generate a set of precise hypotheses concerning behavioural responses and psychological mechanisms and subject them to rigorous tests using data from the real world.

We can ask questions about the history and development of a trait both over geological time (its *phylogenetic cause*) and within the lifetime of an individual (its *ontogenetic cause*), determine how a behaviour enhances survival and reproduction (its *functional* or *ultimate cause*) and identify the factors that trigger a particular behavioural response to occur (its *motivational* or *proximate cause*). Niko Tinbergen, who won the Nobel Prize in 1973 for his work on animal behaviour, pointed out that each of these questions, while appearing very different at face value, is really just a different way of asking the same question – *why* does an animal display a particular trait? – with the answer pitched at different levels of evolutionary explanation. Each of these four senses of 'why' is important, and each can be equally informative. But it is very important not to confuse these levels of explanation by providing, for example, a proximate level answer to a question that asks about the function of a behaviour. Partitioning the kinds of questions we can ask in this way is now known, in his honour, as *Tinbergen's Four Whys*.

By formulating our questions carefully and making sure our answers are pitched at the appropriate level of explanation, we can identify whether behaviours are adaptations produced by the process of natural selection, whether they are by-products of selection for other traits, whether they were initially selected for other purposes but have been co-opted by evolution to serve a new role (sometimes known as 'exaptations') or whether they serve no evolutionary function at all. In other words, the aim of the evolutionary approach is to understand the advantages that traits confer on individual organisms, how these interact with other traits (for example, how having a large brain means that it takes longer for an animal to reach sexual maturity) and how a species' evolutionary history constrains the range of adaptations that are possible.

genetic determinism: the evolutionary red herring

What an evolutionary approach does not involve, however, is any notion that all behaviour is genetically determined and that our biology is our destiny. This issue continues to exercise many people – mainly social scientists, but some biologists have also become surprisingly consumed by it. Much of the criticism levelled at evolutionary approaches to human behaviour seems to rest on the belief that an evolutionary explanation of behaviour necessarily implies that behaviour must be genetically determined. At face value, this may seem a reasonable conclusion to draw. After all, most discussions concerning the evolution of behaviour are explicitly couched in terms of 'the gene for a behaviour'; moreover, the success of a given behaviour is explicitly measured in terms of its *fitness* (a term from population genetics that refers to the relative number of copies of a particular gene that an individual contributes to future generations).

Given this, it might indeed seem to follow that any discussion of evolution must mean genetic evolution. The logic of this argument would appear to be inescapable. But the fundamental question we have to ask is: does it have anything to do with the evolutionary study of behaviour? The short answer is no. There is a world of difference between claiming that we can provide an evolutionary explanation for behaviour and claiming that we are offering an explanation in terms of the genetic determination of behaviour. This is so for two reasons. First, no known species of organism (with the possible exception of single-celled creatures like viruses and bacteria) shows genetically determined behaviour in this way. Behaviour is simply too complex to be determined by single genes. More importantly, if a behaviour truly were genetically determined, it would mean that the behaviour always developed in exactly the same way in each individual and that environmental influences exerted no influence whatsoever. This would result in behaviour that, by necessity, would be completely inflexible: the organism would always behave in the same way, irrespective of the circumstances. Genetic determinism on this scale is an excellent recipe for the rapid extinction of the species in question; it is not a particularly helpful foundation on which to base an effective interaction with a complex, constantly changing world.

Vertebrates evolved large brains precisely to allow them to adjust their behaviour to suit the circumstances in which they happened to find themselves on a moment-by-moment basis. The genes that code for the brain have been selected expressly to enable the organism to escape from a genetically driven existence. Ironically, given the fears of genetic determinism and the loss of 'free will', it is our genes that free us from these deterministic constraints.

An evolutionary approach to understanding behaviour is most definitely not about identifying a single causal link between genes and behaviour. This misunderstanding often arises because an evolutionary approach does require some genes in the system, so convention enjoins us to identify some arbitrary notional gene as the focus for our thinking. The genes in evolutionary explanations are no more than a device for keeping our thinking straight. This does not necessarily mean that there are no specific genes involved, of course, but that is a question that has a purely empirical answer, which must be provided by developmental biologists, not by evolutionary psychologists.

Second, the evolutionary study of behaviour is not actually about the genes that determine *behaviour*, even in the weak sense that there must always be *some* genetic constraints on the capacity to behave at all. The point is that an evolutionary approach is concerned with a *strategic* analysis of behaviour: *why* does the individual behave in this way, in the sense of 'what purpose does it serve for the individual?'. A strategic view makes no specific assumptions about what determines behaviour, it simply assumes that an individual's choice of behavioural strategy is guided by evolutionary considerations (that is, maximising its contribution to the species' gene pool in future generations).

darwin, genes and behaviour

The evolutionary approach to the study of behaviour raises four separate points that need to be clarified:

First, such explanations sound as though (and have certainly been interpreted as implying that) animals make explicitly conscious decisions about their genetic future. No organism can do that, not even humans. Rather, this kind of explanation makes no assumptions at all about how such decisions are made: it could be entirely genetically driven and unthinking, but it could equally be entirely learned and deliberate, or it could be anywhere in between. Which of these possibilities is correct is an interesting empirical question but the answer does not have any implications for whether animals are behaving strategically, or, indeed, whether evolutionary considerations have had a hand in their decisions.

Second, while organisms which behave in a way that increases the number of their descendants in future generations can be considered to have higher fitness, this does not mean that the actual goal of that behaviour is the maximization of fitness. The goal of an Ache hunter from Patagonia may be, on one occasion, to hunt and kill a tapir, or on another to marry off one of his children and dance at the wedding. The link to fitness can occur very far down the line and there is no reason to expect people, any more than other animals, to show behaviours that are overtly designed to increase their fitness (the number of descendents they leave), even though that is their eventual consequence. The achievement of a much more proximate goal can have fitness-enhancing effects, but there need be no direct link between the two. This extended link, via a series of intermediate proximate goals, between behaviour and its ultimate fitness consequences, allows us to explore organisms' behavioural decisions by focusing on immediate short-term consequences such as maximizing energy intake (in the case of hunters) or maximizing the number of offspring sired (in the case of mating strategies), while assuming that successful solutions to these proximate problems will eventually carry through into higher fitness. In behavioural ecology, this is known as the *phenotypic gambit*.

Third, the assumption that organisms are designed to behave in such a way as to maximise their genetic fitness is a heuristic device rather than a presumption of fact: it provides us with very precise predictions, which can be subjected to clear empirical tests. In contrast, the criticism of genetic determinism is explicitly focused on the machinery that permits behaviour to occur – in effect, what enables the hardware to be produced. This is a *how* question and is clearly entirely different from asking *why* behaviour occurs.

Fourth, evolutionary explanations are statistical. Perhaps the commonest attempt to counter an evolutionary explanation is: 'Well, my children don't do that!'. A specific example, however, cannot negate a statistical rule. To disprove the claim, you need to show that *on average* children do not behave in this way. The statistical nature of evolutionary explanations is important – indeed crucial – because evolutionary change cannot happen if everyone behaves in the same way. Organisms have to constantly test their environment, whether this be physical or social, in order to determine whether they are behaving in an evolutionarily optimal fashion. Some individuals will inevitably get it wrong. But, now and again, this trial and error learning will yield a novel solution that is better than all the others. Gradually, this solution will spread through the population, as those who have it (or adopt it) reproduce more successfully. But even so, that solution will never be adopted by everyone in the population: individuals will continue to try out new ones, and some will continue to get it wrong.

In short, the dispute confuses two quite different kinds of question that one might ask of the world: *why* something occurs or *how* it occurs. The confusion probably arises because the word *gene* is used in both kinds of explanation. One focuses on genes as *causes* of behaviour (or the capacity to behave), the other focuses on genes as *consequences* of behaviour (that is to say, the effect that behaving in a particular way has on the genetic make-up of the next generation). Although evolutionary biologists keep these two meanings clearly separated in their minds, those who are less familiar with this approach often confuse them.

Although these two processes are necessarily linked, it does not follow that, in any particular case, the same set of genes is both cause and consequence. In large-brained organisms like mammals and birds, this evolutionary loop is often closed by the brain. Consider an organism that has a large brain, which enables it to adopt flexible behavioural strategies. This allows it to fine-tune its behaviour, in the light of current circumstances, so as to maximize the number of matings it achieves, thereby maximizing the number of offspring it contributes to the next generation. What is passed on from generation to generation and so makes both evolution and the behaviour possible, are the genes for a big brain. But the genes that code for the brain do not *determine* the behaviour (mating) that the brain gives rise to; rather, they merely determine the capacity to make flexible decisions that are well tuned to local circumstances.

Finally, it is worth remembering that when Darwin first formulated his theory of natural selection, he had no knowledge of genes at all. In fact, his new theory was much criticized for containing what many regarded as a very inadequate mechanism of inheritance. Darwin's theory of evolution by natural selection was only rescued from the growing obscurity into which it fell after his death by the rediscovery of Mendel's laws of inheritance.

Although Gregor Mendel, abbot of the monastery at Brno (in what is now the Czech Republic), was developing his laws of inheritance at the same time as Darwin was developing his grand theory, his ideas were not widely appreciated outside his home town (Darwin, who had a copy of Mendel's paper, certainly failed to understand their significance). Remarkably, this key which unlocked Darwin's grand theory remained overlooked in the dusty volumes of obscure libraries for more than half a century until it was rediscovered by geneticists in the early 1900s. The result was what is known today as the *new synthesis* – the amalgamation of Darwin's theory of evolution by natural selection and Mendel's laws of inheritance into a single unified theory.

In any case, Mendel didn't know about genes either! For both Darwin and Mendel, inheritance was all about 'fidelity of copying' between parents and offspring. This has one very important implication: evolutionary processes do not have to depend on genes. *Anything* that causes a correlation between parents and offspring has the capacity to be a Darwinian process. The things that an organism learns in its lifetime and passes on to its offspring can also undergo a process of natural selection. It is entirely possible and equally evolutionary, for non-genetic inheritance to take place and for such non-genetic resources to be selected over time. Cultural processes can therefore have very important evolutionary effects and this is especially true of our own evolution. In other words, understanding human behaviour from an evolutionary perspective may not require the involvement of any genes at all.

disentangling the web

In this book we will rely heavily on a strategic perspective. At each step, we will ask how humans behave in some particular respect. We will then go on to ask what cognitive and physiological mechanisms underpin this behaviour. Where we can, we will ask about the developmental processes involved, in an attempt to address the question of how genetic inheritance and learning interact to bring such behaviour about (however, what we can do in this respect is presently severely limited by our almost total ignorance of the processes involved). And, finally, in a few cases, we will ask questions about the evolutionary history of a particular phenomenon (although the number of cases where we really can say anything useful about this is even fewer). For the moment, our concern will mainly be to raise questions about the processes involved and point to possible ways ahead.

Evolutionary psychology has often been seen as an alternative to more conventional approaches in psychology, the equivalent of developmental psychology, cognitive psychology or social psychology. That, however, is to misunderstand what the evolutionary approach is all about. In biology, the evolutionary approach provides a unifying framework that allows different subdisciplines (behaviour, ecology, physiology, genetics, anatomy, biochemistry, etc.) to talk to each other. In effect, *Tinbergen's Four Whys* spell out how the various subdisciplines are related and allow them to interact without confusing the issues or getting into pointless disputes. In our view, evolutionary psychology supplies the same service for psychology, creating a theoretical framework for unifying the various subdisciplines. To all intents and purposes, functional questions about why individuals behave the way they do (known in biology as behavioural ecology) are really just social psychology with an evolutionary backbone. Cognitive and developmental psychology, in turn, map neatly on to the mechanism and ontogenetic senses of *why?*.

Only evolutionary history (phylogeny) is missing from conventional psychology. Despite Darwin's interest in the evolution of the mind, psychologists have tended not to ask questions about the evolutionary past, instead, they have taken the present as the focus of their interests. But there is a good reason why psychology should be interested in evolutionary history. Comparative psychology has always stood as a reminder to psychologists that we share our evolutionary past with other animals and in particular with the primates. Understanding just how and why we differ from non-human animals is a psychologically interesting question and knowing when those differences emerged may provide us with important insights into human nature.

In this book, we will not have much to say about the behaviour and psychology of non-human animals, even though comparative psychology is an important branch of evolutionary psychology. This is simply because it would require a much longer book to bring it all together. Animal research will, none the less, constantly be hovering in the wings, not least because almost all theories of behavioural ecology were developed through studying animals. The functional side of human evolutionary psychology thus builds on a vast mass of research: in applying these ideas to humans, we ask to what extent the same general principles underpin human behavioural decisions.

By the same token, we will have little to say about many of the more conventional aspects of cognitive psychology such as memory, perception, thinking and so on. These largely focus on questions about mechanisms, the fundamental building blocks of how we interface with the world. They are there and they surely have an evolutionary origin, but, consistent with our focus on strategic functional questions about behavioural decisions, our main concern will be with what has become known as *social cognition*, a higher-order layer of cognitive mechanisms specifically involved in the social decision-making that lies at the heart of human behaviour.

One last source of confusion needs to be clarified. Those who apply an evolutionary approach to human behaviour have, for the past decade or so, been locked in a trenchant and, at times, rather unseemly dispute about how such studies should be done. On the one side, those with a background in biology (and specifically, behavioural ecology) have stressed the importance of asking whether behaviour is adaptive in the conventional functional sense used by biologists (that is, that a particular behaviour has the consequence of enabling the organism to maximize its fitness). They have emphasized both individual differences in behaviour and the analysis of their functional consequences.

In contrast, those with a background in psychology have tended to focus on the universals of behaviour that are true of the species as a whole. As a result, they have concentrated on the cognitive mechanisms that produce behaviour – the *design* of the human mind, as it were. Because they view the neuro-cognitive hardware rather than behaviour as being the product of selection, they have insisted that the behavioural ecology approach is, in the case of humans, rather fruitless: they argue that there has been little change in the human gene pool in the last 10,000 years, since the invention of agriculture, and hence that much of our behaviour will inevitably be maladaptive because we are stuck with a Stone Age mindset in a modern industrial environment. The human mind, they argue, evolved to deal with conditions in the Environment of Evolutionary Adaptedness (or EEA), the historical time and place in which our ancestors evolved their particular characteristics. Hence we can only understand the mind as an adaptation if we see it against the background of the prehistoric environment in which it evolved.

We see little or no benefit in polarizing an artificial distinction. The fact is that the human mind, like the minds of all species with brains of a decent size, evolved to cope with environmental variability. The terrestrial environment has never been stable at any time in the Earth's long history and any species that sought to evolve rigid cognitive mechanisms would be signing its own extinction warrant. Most vertebrate species are designed to be smart precisely so that they can adjust their behaviour to the constraints of current circumstances, whatever they happen to be. This is not, of course, to deny that some aspects of the human mind may be much less flexible than others. Rather, it is to say that we should not be prescriptive about what these might be until we have shown that they actually exist.

Instead of getting involved in an argument that is ultimately pointless and distracting, we prefer to bring both perspectives together as best we can. Cognition is an essential element in any account of the functional aspects of human behaviour. While it remains true that the structures of the human mind evolved in a particular environment, the EEA is an elusive concept since our minds, like our bodies, are the product of a long evolutionary history and it is probably not possible to identify a single point at which any one feature came to be.

In the chapters that follow, we will present evidence to show that many aspects of modern human behaviour are functionally adapted to evolutionary goals and that behavioural plasticity and flexible decision-making are key to achieving these goals. At the same time, we will also find aspects of human behaviour that seem to be resolutely intransigent in the face of changing environments. As a result, we will need to develop an integrated approach that draws together a whole range of disciplines to understand the complex explanatory web which underpins the way humans behave.

summary

An evolutionary approach provides us with a powerful framework for studying human behaviour and the mind. This is not because it offers us a different method to conventional psychological approaches but because it allows us to integrate them under a single unifying theory; Darwin's theory of evolution by natural selection. In doing so, it is important to remember that an evolutionary approach does not necessarily imply that either behaviour or the mind that underpins the behaviour is in any way genetically determined. Learning is itself a Darwinian process and provides one of several possible alternative mechanisms of inheritance in addition to conventional genetic processes. Appreciating this enables us to widen the scope of things we study to include culture and the mechanisms of cultural inheritance.

what evolution did for us

When Darwin first developed his theory, people assumed that animals behaved in ways that were good for the species as a whole. For example, female lionesses which suckled young cubs belonging to other females in their pride were assumed to be doing so in order to make sure that there were plenty of lions in the next generation and so the species wouldn't become extinct. However, the most important thing to note about the theory of natural selection is that it is concerned with *individual* survival and not with the survival of the species. Although individual reproduction inevitably has the effect of perpetuating species, this in itself is not the purpose of reproduction (or evolution).

Individuals are selected to behave in their own reproductive interests and the fate of the species as a whole is irrelevant to individuals' reproductive decisions. This must obviously be the case if natural selection is to operate in the way Darwin envisaged: since the whole process is based on the notion of inter-individual competition, any organism that behaves so as to benefit the species or group at some cost to its own reproductive interests is likely to leave fewer descendants than less noble-spirited individuals who just look after themselves.

darwin and natural selection

So how did Darwin envisage natural selection as operating? While his views on the importance of natural selection in the evolutionary process changed over the course of his lifetime and evolutionary biologists today continue to argue over the relative importance of selection as a means of evolutionary change, there is no doubt whatsoever that, with this idea, Darwin changed forever the way we think about the natural world.

The theory of natural selection is deceptively simple and is based on three premises and their logical conclusion:

Premise 1: All individuals of a particular species show variation in their behavioural, morphological and/or physiological traits (their *phenotype*). This is usually known as the *Principle of Variation.*

Premise 2: A part of this variation between individuals is *heritable*: that is, some of that variation will be passed on from one generation to the next (or to put it even more simply, offspring will tend to resemble their parents more than they do other individuals in the population) – the *Principle of Inheritance.*

Premise 3: Whenever there is competition among individuals for scarce resources such as food, mates and somewhere to live, some of these variations will allow their bearers to compete more effectively than others. This competition occurs because organisms have a capacity to greatly increase in numbers and produce far more offspring than can ever give rise to breeding individuals (just think of frogspawn, for example) – the *Principle of Adaptation.*

Consequence: As a result of being more effective competitors, some individuals will leave more offspring than others, because the particular traits they possess give them some sort of edge: they are more successful at finding food, or mating, or avoiding predators. The offspring of such individuals will inherit these successful traits from their parents and 'natural selection' can be said to have taken place. Through this process, organisms become adapted to their environment. The success with which a trait is propagated in future gener-ations, relative to other variants of that trait, is called its *fitness.* Fitness is a measure of relative reproductive success (that is, relative to alternative variants of the same trait); strictly speaking, it is a property of traits. This is sometimes known as the *Principle of Evolution.*

By specifying a mechanism by which evolutionary change could be effected, it then became possible to formulate testable hypotheses aimed at explaining the anatomy and behaviour of organisms. If a trait was an adaptation, then it should show evidence of being well adapted to the purpose it was supposed to serve; and if it continued to confer a selective advantage on the organism that possessed it, then it should also help to increase the survival and reproductive success of those organisms relative to those that did not possess it (or which possessed inferior versions of it).

A second important consequence of Darwin's position was that it made 'group selection' (evolution for the benefit of the species) an extremely unlikely (though not entirely impossible) explanation for the evolution of anything. Despite this, group selection remained firmly ensconced in the public imagination. Indeed, even biologists often failed to appreciate this point and it was not until the 1960s that the concept of group selection was finally laid to rest. Evolutionary biologists have remained extremely cautious of mentioning group selection ever since.

the 'selfish gene' as shorthand

Sometimes, however, even the individual is too gross a level to understand the workings of evolution. This is because, although natural selection acts on the survival and reproductive success of individuals, what actually changes over time is the frequency of genes in the population's gene pool. Individuals are really transient beings: no matter how long their lifespan, they all die in the end. Genes are the entities that persist and provide continuity over time.

In his famous book *The Selfish Gene*, Richard Dawkins argued that there are some aspects of evolutionary biology which we can understand much better if we adopt a gene's-eye view of the world and recognise that the evolutionary process consists of genes which help to promote the survival and reproductive success of the bodies in which they find themselves, rather than vice versa. To get this idea across more clearly, Dawkins made a distinction between 'replicators' and 'vehicles'. Replicators are the entities (genes) that reproduce themselves and persist through time, whilst 'vehicles' are the entities (bodies) that the replicators construct to contain themselves and which increase the replicators' ability to reproduce and leave as many descendants as possible.

For supposedly advocating 'genetic determinism', Dawkins came in for a lot of misguided abuse, mostly from people who didn't take the trouble to find out what he really meant (see Malik [2000] for a review). It is vital to appreciate that when Dawkins talks about genes in this way, he is not suggesting that individual genes are consciously striving for their own ends; it is simply a shorthand way of speaking about evolutionary processes. What it really means is that, all else being equal, animals whose genes lead to the development of traits that increase an individual's ability to survive and reproduce are more likely to be represented in the gene pool in succeeding generations than are individuals who had a different array of genes that resulted in traits that weren't so successful in that particular environment. That is such a mouthful that no sensible biologist would want to repeat it every time he or she wanted to discuss the evolution of something. Dawkins provided us with a convenient contraction that needs only two words. However, it *is* important that, when we use his phrase, we bear in mind that it stands for that over-long sentence, and nothing more.

The significant point is not that a particular gene causes a particular behaviour, but that genetic differences between individuals (whatever these may be) are linked to behavioural differences that, in turn, result in certain individuals being more reproductively successful than others. That, as we noted in Chapter 1, is how genetic fitness is defined. Natural selection is always about relative differences between individuals, not absolute ones.

We also need to remember that evolution is always something of a compromise: at any one time, there are numerous selection pressures acting on the individual, in many different ways, with the result that a given adaptation may not always be the perfect solution to the problem in question. The classic example is that adaptations designed to enhance reproductive capacity are inevitably compromised by those geared toward enhancing survival. For example, a male could have enormously high fitness if he did nothing but mate all day but his mating activities are likely to be curtailed prematurely if he doesn't spend some time feeding. Generally speaking, most organisms are jacks-of-all-trades and masters of none. In this sense, individual organisms, and not genes alone, are the units of selection, since the process of natural selection acts on the organism as a whole and not on genes in isolation.

It is also important to remember that other evolutionary processes can shape traits. Some traits may be historical accidents, produced by a sudden reduction in population size, such that only a very few individuals leave descendants from which the population can build up again. These *founder effects* can result in traits being fixed in populations despite the fact that they confer no real benefit on their holders – and may, in some cases, be detrimental. In a similar way, developmental constraints may result in traits that have not been directly selected for, but which have 'come along for the ride' as a consequence of selection for other traits.

Space does not permit a full review of all these alternative evolutionary mechanisms, but suffice it to say that, when attempting an evolutionary analysis, we must be very careful to exclude all other possible explanations for a trait before accepting that something is an adaptation. Equally, we must not be quick to dismiss something as an adaptation merely because its evolutionary function is not obvious. Doing so almost always reflects a lack of knowledge on our part. Prematurely concluding that a phenomenon has no adaptive function is as heinous a sin as prematurely concluding that it does.

altruism and the gene's-eye view

A gene-centred perspective on behaviour has been viewed as somewhat reductionist, as attempting to reduce something as complex as behaviour to something that is much simpler, like genes. However, when we take a 'gene's eye view' this is not to imply that behaviour is genetically determined. As we explained in Chapter 1, all behaviour is the result of an interaction between genes and environment and, in the next chapter, we go into this in more detail in order to emphasize that gene–environment interactions are the key to understanding how behaviour develops in an organism.

A gene's eye view has been of great theoretical value, since it has given us a way to understand certain facets of animal behaviour that were otherwise puzzling, since they seemed to require a 'good for the species' argument that didn't quite square with Darwin's theory of natural selection. Consider the female lions we mentioned at the beginning of this chapter: if behaviour must always be to the advantage of the individual rather than the species, why should a lioness suckle other mothers' cubs and help promote their genes at the expense of her own? Such behaviour, where one animal provides a benefit to another, at a cost to itself, is termed *altruism* (which means 'being unselfish') and was one of the puzzles that taxed Darwin himself when he was developing his theory of natural selection.

Sadly for Darwin, this problem wasn't solved until 1964, when a young graduate student, W.D. (Bill) Hamilton, pointed out that altruistic behaviour could evolve if the individuals that benefited from the behaviour were related to the altruist. This is because close relatives share some of their genes in common: two siblings share roughly 50 per cent of their genes, while two cousins share 12.5 per cent, which they inherit from a common ancestor (parent or grandparent, respectively). If a female lion possesses a suite of genes that cause her to help raise her sister's cubs, there is a good chance that the genes in question will be passed on, even if that female has no offspring of her own. This is because her sister has a 50 per cent chance of having inherited an identical copy of those genes, which she then passes on to her offspring. As far as evolution is concerned, it doesn't much matter whose body the genes are in, as long as they get passed on.

So, the reason that lionesses in a pride feed one another's cubs is because they are all sisters. Far from behaving unselfishly, female lions are actually helping themselves – or more exactly, their genes – by helping other animals. This kind of process, where animals help promote the survival and reproductive success of their relatives, is known as *kin selection*.

However, altruistic behaviour can also occur between animals that are not related to each other, and so kin selection cannot explain all cases of altruism. An alternative explanation for co-operation under these conditions comes from Robert Trivers, an American evolutionary biologist. He argued that it would be an advantage for animals to help non-relatives if they could be sure that the favour would be repaid at a later date. In this way, the benefits balance out. Obviously, this only works if animals interact with the same individual on a number of occasions (so that the benefits are swapped fairly) and also if they are able to recognize one other.

This process is known as *reciprocal altruism* and, compared to kin selection, occurs much more rarely. This is because, when benefits are exchanged in this way, there is a delay between one animal giving the benefit and the other returning it, which makes it rather easy for the second animal to cheat and take the benefit without repaying it. If reciprocation is not reliable, then it is not in the first animal's genetic interest to co-operate at all and the exchange of benefits can never get off the ground. Behaviour systems based on reciprocal altruism are therefore much harder to get going than those based on kin selection.

A third evolutionary explanation for altruism, or co-operation, is known as *mutualism*. In this case, two or more animals co-operate to achieve a goal that benefits both of them at the same time. Co-operative hunting is an example of this: two animals that co-operate may be able to kill a much larger prey than either would be able to on its own. Co-operating lions, for example, are able to bring down zebra or buffalo, whereas lions hunting on their own usually have to make do with antelope, which weigh only a fraction of a zebra or buffalo.

is the gene's-eye view too narrow?

The mechanisms of kin selection, reciprocal altruism and mutualism form the basis of most evolutionary explanations of altruism and it has become something of a heresy to suggest that altruism can evolve as a result of selection at any level higher than the individual. However, for many years, the evolutionary biologist David Sloan Wilson has argued against this position. Along with the philosopher Eliot Sober, Wilson has pointed out that, whilst group selection is indeed unlikely to occur, it is by no means impossible for selection to take place *at the level of* the group.

One striking example of this is that, within our bodies, our genes do not aggressively compete with each other for chromosome space, but co-operate in their collective replication and transmission. Similarly, we have many cell types living in harmony: individual cells do not behave entirely 'selfishly', reproducing at the expense of others, because to do so would be to put the vehicle that carries them at risk. (Cells proliferating out of control in this way are what we generally refer to as cancer, with obvious detrimental consequences for the body/vehicle.) Instead, the body's cells are prudent, reigning in reproduction and co-operating for the good of the whole organism.

At the cellular level, selection is at the level of the group (that is, the assemblage of cells that make up an individual) since this enables a functional vehicle to be formed, which can pass on genes to future generations more effectively than can a collection of selfish individual cells. Sober and Wilson argue that, under certain conditions, it is possible for animal groups to function as the vehicles of selection, where the animals that make up those groups evolve traits that help increase the survival of the group at the expense of other groups or individuals.

One of the most crucial conditions to be met is that there must both be competition between individuals in different groups and competition between individuals in the same group. In both cases, we have competition between individuals, as required by the theory of natural selection, but the difference lies in the level at which that competition occurs. Within groups, individuals are the vehicles and they are in direct, selfish, competition with each other. However, between groups, individuals in one group are joined together in the face of competition with individuals in another group. Sober and Wilson argue that, in this case, it is appropriate to consider the group as the vehicle of selection since, under these conditions, certain altruistic traits will be selected in individuals which increase the competitive ability (and hence the fitness) of the group as a whole, even though such traits may reduce the fitness of some individuals in that group, relative to others. That is, although selfish individuals will, on average, have more offspring than altruistic ones, the groups with greater proportions of altruists will produce more offspring in total than those with a greater proportion of selfish individuals (due to the benefits provided by the altruists) and so the total number of altruists will remain stable in the population as a whole.

For this to be true, it is crucially important that groups are in competition with each other and not isolated, each living on its own ecological island. If there were no compensating advantage to the group as a whole, then altruistic individuals would be ruthlessly exploited by selfish individuals and natural selection would soon eliminate all the altruists, who would have no way of bringing pressure to bear on those who tried to exploit their generosity, since the selfish exploiters wouldn't care much whether or not the group existed. Groups also need to periodically split and re-form into new combinations, or at least show fragmentation and movement of individuals between groups, in order to prevent non-altruists from eventually coming to dominate a particular group.

Sober and Wilson refer to this as *multi-level selection theory* (MST) and call their specific mechanism of selection *trait group selection*, since a group can be defined as the set of individuals sharing a particular trait, as well as a set of individuals forming a cohesive group, as we would usually think of it. As an example of this kind of effect, Sober and Wilson describe how chicken-breeding programs in America, designed to increase the egg productivity of hens, have frequently produced strains of hyper-aggressive chickens that have a lower productivity than their progenitors. This is because, in today's intensive poultry industry, competing aggressively for food and space may be an important factor in influencing whether a hen is a good egg layer. However, breeding selectively from these individuals may produce a population that is so aggressive that the resulting stress inhibits their laying. But, if the most *social* females are selected and placed into new groups, it is possible to increase egg-productivity by 160 per cent – a figure far in excess of that produced by standard individual-based breeding programmes. This is because, as well as selecting for egg-laying abilities, the breeders are selecting for traits that allow females to function well in groups. Thus, instead of aggressive chickens that are poor layers, selection at the group level produces chickens that lay more eggs per day, have lower mortality and show so little aggression that they do not need to have their beaks trimmed to prevent them pecking other hens.

By no means all biologists accept Sober and Wilson's argument, since many insist that the precise conditions required for it to work would be very unlikely to occur in the real world. This, of course, is an empirical issue and we cannot at present say whether or not this is the case. Other criticisms, however, are less valid, since they assume that Sober and Wilson are arguing for the kind of 'good for the species' group selection that biologists have shown to be biologically implausible. It should be obvious that MST differs from this kind of group selection, since conventional individual selection is an integral part of the Sober–Wilson theory. Traits that evolve at a group level do so because, ultimately, they bring fitness advantages to the individuals which make up those groups. Selection for individual traits at the group level is taking place and this then enables some groups to do better than others. Identifying strongly with one's own group and showing a willingness to aid another individual on that basis alone is a trait that is selected for in individuals, but it can only operate at the group level. If nothing else, without the presence of groups, it wouldn't be possible to discriminate against non-members, or to behave differently towards them; the selection pressure to do so simply wouldn't exist.

Wilson argues that MST may be particularly relevant to humans, precisely because we are so intensely social. Individuals that got along better with their peers would have tended to leave more descendants – whether because their stress levels were lower, their mortality was reduced since they suffered less aggression from others or because sociable individuals and their offspring were more likely to receive assistance from others than were more irascible members of the group. If our skills at dealing with others were selected in just this kind of group context, it may account for our abilities to co-operate with people we have never met before (or that we never meet at all in today's Internet age). Taking multi-level selection seriously would therefore seem to be essential when trying to provide an evolutionary explanation for many of our most striking social traits, and we shall return to it again throughout this book.

Interestingly, altruism is one of the few areas that has been given equal attention by evolutionary biologists and social psychologists. Psychologists have been interested in identifying the factors that cause individuals to act co-operatively and help one another altruistically. The in-group effect we described above, whereby people show a strong tendency to identify with and help others whom they perceive as belonging to the same group, is an extremely robust phenomenon, and has been studied by social psychologists for more than half a century. It can be produced even when groups are designated on the basis of quite arbitrary criteria (being allocated one geometric shape versus another or on the basis of preferences for paintings by Picasso over those by Matisse).

These psychological traits are the kinds of things that might be selected for at the level of the group and could help give one group a competitive advantage over another by increasing the cohesiveness of the group and making individual members more willing to defend and protect their group mates. Such traits can also help increase group harmony overall, so that the stresses and tensions of group living are reduced, which in turn can increase the relative reproductive success of a cohesive group over a fragmented one.

niche construction theory

One last evolutionary theory that we need to mention before we start reviewing human psychology and behaviour in earnest is Niche Construction Theory (NCT). Like multi-level selection, NCT is particularly pertinent to issues of human evolution, both anatomical and psychological. John Odling-Smee, who has been working on the ideas behind this theory for many years, coined the phrase 'niche construction' to get across the central point that animals do not passively occupy ecological niches but actively modify them.

Active modification of an ecological niche by an organism changes the selection pressures that act on the organism itself: in effect, individual organisms may become the engineers of their own evolution. Spider's webs, for example, modify the selective environments of the spiders that spin them, creating new opportunities for selection to act. Other forms of niche construction modify the selective environment of the constructing organism's descendants. For example, there are many insects that provide their eggs with food. They may lay their eggs on a leaf, or even, in the case of parasitoids, in the body of another organism. In such cases, the modified niche is an example of what Odling-Smee and his colleagues, Kevin Laland and Marcus Feldman, call 'ecological inheritance'. Ecological inheritance can have a profound effect on the evolutionary process since it represents a second form of inheritance that differs from standard genetic inheritance. Inheritance of land, chattels, money and status play an especially important role in human societies and may thus represent a particularly dramatic example of this process.

In other words, ancestral organisms can also transmit phenotypically modified habitats to their descendants, as well as their genes. If these ecologically inherited niches remain stable over time (that is, the process of ecological inheritance persists across generations) then they will result in new selection pressures being applied to organisms and new forms of adaptation arising, which may then lead to further modification of the niche by the organism. This, in turn, implies that environments can evolve as well as organisms.

Niche construction, then, is essentially a feedback process, and it is this feedback which gives it its evolutionary significance. Theoretical analyses using population genetic models have shown that traits whose fitness is affected by niche construction (so-called recipient traits) co-evolve with the niche-constructing traits themselves. In our own history, for example, the evolution of stone tools (a niche constructing trait) expanded the range of foods that early humans were able to eat (to include meat and bone marrow), thus selecting for changes in our digestive morphology and relieving a constraint on the evolution of brain size. Tool use, dietary changes and brain size continued to co-evolve and feed back on each other in evolutionarily significant ways.

Niche construction means that adaptation is no longer a one way process, whereby organisms respond exclusively to environmentally imposed problems but becomes two way, with populations of organisms setting problems for themselves, as well as solving problems set by the environment.

This has important implications for how we view evolution, since it enables experiences that an animal undergoes during its life to have an effect on the evolutionary process. When organisms niche construct, they become more than just 'vehicles for genes' because they are now able to modify the sources of natural selection that are present in the environment and so have some responsibility for selecting their own genes. Moreover, there is no need for niche-constructing activities to be genetically specified. Learning, and other forms of experience, may lead to animal niche construction; in humans, it may also depend on culture.

It should now be clear why niche construction is so relevant to an explanation of human evolutionary ecology and behaviour. We show a more diverse and sophisticated form of culture than any other species on the planet, and we have been constructing our own niche for hundreds of millennia – since, at the very least, the time we first invented tools, around two million years ago. The philosopher Matteo Mameli has argued that other humans may also have played a powerful niche constructing role during the course of human evolution, shaping our psychology and in particular our mind reading abilities – our ability to attribute thoughts, feelings, beliefs and desires to others – so that today, human psychological development is utterly dependent on the presence of other human minds for its normal expression: we are 'mind shapers', as well as mind readers (we will deal with this idea in more detail in Chapter 5).

However, as the philosopher Kim Sterelny points out, the fact that we have been constructing our niche for so long does raise some problems for understanding the evolution of human cognition, because it means that humans have, to some extent, freed themselves from the constraints imposed by the environment. Thus, while we might attempt to reconstruct the ecology of a species from a know-ledge of habitat, weather conditions, predator densities and the like, this may be much more difficult when trying to understand patterns of human evolution, because much of our evolutionary history has been spent constructing our own niche, rather than being shaped by independent features of the natural environment. The selective envir-onment of humans may therefore have been very changeable, even during periods when the physical environment remained entirely stable. For example, once hominids invented a means of carrying water with them, they were freed from the selection pressures imposed by increasing aridity in the physical environment. If this happened on a small local scale, it would leave few traces in the fossil record and make it difficult to determine the exact evolutionary course that humans had taken at this point.

Sterelny therefore suggests that we need to use a diversity of methods to probe the evolution of human cognition, including experimentation, modelling and computation, comparative studies of other species, archaeology and task analysis (where the cognitive demands of particular tasks are identified). It is also important to understand the adaptiveness of current behaviour, since this helps to reveal how our niche-constructing abilities influence the behavioural strategies that people follow, the cues people use to guide these strategies and the plasticity that people can display in the face of environmental constraints. Combining behavioural studies of humans with experimental psychology, along with the historical sciences of palaeoanthropology and archaeology, allows us to constrain the degrees of freedom we have in constructing a plausible scenario and moves us from 'Just-so' story-telling to hypothesis formation and testing. This is by no means an easy task and we are still very far from having achieved it. Conversely, the fact that the situation is more complex than we might originally have imagined does not, of itself, make the task impossible.

the human revolution

This brings us to one final issue – the history of human evolution. This is important to our discussion for two reasons. One is that, building on Tinbergen's *Four Whys*, understanding the historical origins of our behaviour and psychology may help us appreciate their functional (or adaptive) significance. The other is that a particular view of human evolutionary history has come to occupy a more prominent position in evolutionary psychology than it really deserves.

Our lineage, the hominid (or in some terminologies, hominin) lineage, is a member of the African Great Ape clade (or family). Indeed, we share a more recent common ancestor with the chimpanzees than either of us shares with the other two Great Apes, (the gorilla and orang-utan). According to the genetic evidence, the human and chimpanzee lineages separated some time around 5–7 million years ago (MYA). Since there is little fossil material from that period and what little there is is controversial, we are not able to say very much about this period of our history other than that we come from a fairly typical Great Ape line. The earliest known members of the hominid lineage for which there is plentiful fossil evidence

(the australopiths of the genera *Australopithecus* and *Paranthropus*)
are in many ways (but especially in terms of brain size) fairly standard apes. They differed from our Great Ape cousins only in that
they walked bipedally, whereas apes normally walk quadrupedally
(on four legs).

The big change came around 2.5 MYA, with the emergence of the
genus *Homo*, to which modern humans belong. This was marked by
an expansion in brain volume (from about 400cm^3 to about 650cm^3
– still a long way short of the 1350cm^3 typical of modern humans), a
rapid increase in stature, some significant changes in the anatomy of
the legs and hips (allowing more fluent bipedal striding) and, perhaps most significant of all, increasingly sophisticated stone tools.
Although the later australopiths had probably begun to develop
stone tools, these tended to be relatively crude hammers. With the
appearance of *Homo ergaster* (around 2 MYA), stone tool manufac-
ture underwent a dramatic shift, into what is known as the
Acheulian industry, which is associated with the production of large,
carefully made, symmetrical, tear-shaped hand axes.

The *Homo ergaster* period is accompanied by a number of important ecological changes in lifestyle. These included the occupation of
more open (as opposed to wooded) savannah habitats further away
from standing water, a larger ranging area and a more nomadic way
of life. These changes resulted in the same species occupying virtually
the whole of sub-Saharan Africa (except, probably, the densely
forested areas in west and central Africa) and, for the first time in
hominid lineage history, escaping the confines of Africa to colonize
southern Europe and Asia as far east as China. The Asian branch of
this lineage is usually known as *Homo erectus*, although the anatomical differences between the two species are somewhat arguable.

Two points are important about this phase of our history. One is
that we were not alone. Through much of the period during which
H. ergaster/erectus was present in Africa, there were other hominids
alongside it. There may have been as many as five species of australopiths and *Homo* alive at the same time, often occupying the same
habitats. The tree of human evolution is more like a bush than the
traditional view of single straight stem leading from ape-like ancestors to modern humans. The second point is that the *ergaster/erectus*
phase is remarkable for its stability over a very long period of time.
For the better part of one and a half million years, there was surprisingly little change in either the anatomy of the species or the kinds of
tools it made.

The origins of modern humans lie in a transition that occurred around half a million years ago. Although *Homo erectus* may have survived in parts of Asia until as recently as 60,000 years ago, it was replaced, in Africa, by one or more species of archaic humans (generically referred to as *Homo heidelbergensis*). These species are characterized by a significant enlargement in brain size (to about 1200cm³), somewhat more sophisticated stone tools and a relatively rapid dispersal through Africa and into Europe (but not Asia). In Europe, they eventually gave rise to the Neanderthals (*Homo neanderthalensis*), who so successfully occupied the Ice Age habitats of Europe until around 28,000 years ago. But in Africa, the archaic humans gave rise to a new, more lightly-built, larger-brained species: anatomically modern humans (*Homo sapiens*), the species to which we belong.

Anatomically modern humans turn out to have an unexpectedly recent origin. Analysis of the DNA of modern humans from around the world suggests that all humans alive today shared a last common ancestor as recently as 200,000 years ago (and possibly as recently as 100,000 years ago). (We won't discuss the evidence for this in any detail here, you can read about it in any modern palaeoanthropology textbook.) We also now know that anatomically modern humans and Neanderthals belonged to different species. This has been confirmed by comparison of DNA extracted from the fossil bones of Neanderthals and Cro-Magnon peoples (the earliest representatives of anatomically modern humans in Europe). While Cro-Magnon DNA is indistinguishable from that of modern humans, that of Neanderthals differs significantly from both.

The appearance of modern humans in Africa is characterized by the simultaneous appearance of a more sophisticated tool technology about 100,000 years ago, including finely made arrow and spear points, blades with razor-sharp edges (known as microliths) and multi-barbed harpoons. These new weapons seem to mark a shift from hunting styles that use thrusting weapons (heavy spears – characteristic of archaic humans, including Neanderthals) to one using projectile weapons (javelin-like spears and bows and arrows). By the time modern humans arrived in Europe (around 40,000 years ago, some time before the Neanderthals became extinct), this technology had blossomed into fully-fledged art – buttons, beads, needles, Venus figurines, cave paintings and deliberate burials complete with grave goods (the latter from about 20,000 years ago).

The important lesson that has been learned during the last thirty or so years is that human evolution has been far from straightforward. Indeed, at times it hung on a demographic knife-edge. The appearance of anatomically modern humans, for example, seems to be associated with a genetic bottleneck: all living humans are descended from around five thousand breeding females who lived 150,000–200,000 years ago in Africa. They need not all have lived in the same place at the same time; nor need they have been the only breeding females then alive. This means that the total human breeding population (the 'effective population size' as it is called) was once very small indeed. Of all the humans living at that time in Africa, only a very small number (relatively speaking) formed the pool of individuals from which we all descend.

The occurrence of genetic bottlenecks often spells the end for a species, as unlucky accidents with respect to who manages to breed and who doesn't can mean that, for example, congenital diseases are passed on to to all members of the population. More generally, the reduction in genetic variability in the population means that it will be unable to respond adaptively to environmental change, because individuals with the right kind of genetic make-up are missing. Demographic bottlenecks of this kind imply that the species' survival hung by a thread. They are also commonly associated with rapid evolutionary change.

origins of the modern mind

The appearance of Acheulian hand axes seems to mark a significant improvement in cognitive abilities, in particular the ability to imagine the shape of the future axe inside the cobble of raw stone. But even so, the cognitive skills of which *Homo erectus* was capable were evidently not in the same league as those achieved later by archaic humans (including the Neanderthals) or, in particular, anatomically modern humans. These shifts in technological competence imply marked changes in cognition, involving greater foresight, much finer motor control and hand-eye co-ordination, and clear evidence of intention.

One particular feature of Acheulian hand axes stands out. *H. erectus* churned out exactly the same kind of tool for millennia after millennia; those produced at the end of the period were indistinguishable from those produced more than a million years earlier. We only have to contrast this with the speed at which mobile phone technology has changed in the last ten years to see how extraordinary this stability was.

This uniformity suggests that, despite an increase in brain size, *H. ergaster* and *H. erectus* were psychologically very different from modern humans. One suggestion is that they lacked the ability truly to imitate each other. 'True' imitation requires both the ability to understand the intention behind an act and the ability to repeat exactly the behaviours used to produce it. Copying someone's behaviour without really understanding the goal behind it, as, for example, when very young children copy their father shaving, is termed 'mimicry' and is not as cognitively demanding.

True imitation means that, if you notice that the particular technique someone is using is better than yours, you can adopt his or her technique and so improve the quality of your own work. If you then modify the technique and further improve it, others will imitate your technique and take advantage of your innovation; as a result, tool form will gradually change over time. Without true imitation, tool form is destined to remain static.

It has therefore been suggested that, while *Homo ergaster/erectus* individuals would have observed other toolmakers to gain an idea of what a finished tool should look like, when it came to making their own tools, they did it in their own way, with actions that only approximated those of the skilled toolmaker. The result was an idiosyncratic method of tool production, even though the finished product looked the same. Without precise copying, any improvements in tool design would be lost after one generation, since no other individual would be able to replicate precisely the technique used to produce it. If true imitation had been possible then, whenever an improvement occurred, it would have been passed between individuals (most likely from older to younger generations) as each person copied precisely the sequence of actions used by the toolmaker.

One important implication of all this is that there is no one period in the past 5–7 million years which one can point to as a formative phase of human evolution. Our defining traits were acquired piecemeal, over a very long period of time. Bipedalism evolved very early on (perhaps 6–7 MYA); our striding walk and its associated anatomical changes came much later (around 2.5 MYA) but long before the surge in brain evolution that led very rapidly to the massive brains of modern humans (around 0.5 MYA). Meanwhile, although stone tools have a long history (perhaps dating back to 2.5 MYA), there is a series of very distinctive shifts in tool style and quality over time, culminating in the dramatic sea change of the Upper

Palaeolithic Revolution, some time after 100,000 years ago, that suggests a succesion of small but important cognitive developments. The human mind also evolved piecemeal.

This mosaic, characteristic to human evolution, is important, because there has been a strong tendency for evolutionary psychologists to relate aspects of the modern human mind back to the environmental conditions in which it evolved (the so-called Environment of Evolutionary Adaptedness, or EEA). While it may be possible to identify the particular circumstances under which individual components evolved, the palaeontological evidence suggests that there is no generic period when everything characteristic of modern humans evolved as a suite of related and co-evolved traits. We are too much of a hotchpotch. Rather, our traits evolved over an extended period of our evolutionary history, some probably very early on, others very recently indeed.

summary

Evolution does not work for the good of the species, but for the good of individuals. An understanding of the workings of natural selection can often be enhanced by taking a 'gene's eye view', but when doing so, we must always bear in mind that selection acts on the organism as a whole. Taking a 'gene's eye' perspective has helped solve many of the puzzles of evolution but when dealing with our own evolution there are other evolutionary mechanisms that we need to consider. In particular, multi-level selection and niche construction are essential for understanding patterns of human evolution, due to our species' sociality and intelligence. The human line began with a family of upright apes and branched out into an array of specialised and well-adapted hominid species. The modern human suite of physical and behavioural characteristics has been put together slowly over evolutionary time and some of these characteristics were shared with our sister species. For most of our evolution we have not been the only hominid species around and our current domination of the planet is really a lucky accident.

genes, development and instinct

Perhaps the most contentious issue in evolutionary psychology and one we raised right at the beginning of the book, turns on whether it is our genes or our upbringing that determine our behaviour. Some take the Jesuit view, which famously (or perhaps infamously) claims: 'Give me the child until he is seven, and I will give you the man', believing that early childhood is the time when personalities are formed. In the iconic imagery of the eighteenth century empiricist philosopher John Locke, children's minds are 'blank slates' on which life's experiences etch personality and styles of behaviour that then remain stable into adulthood. In contrast, others have argued that 'genes will out' and that, no matter what the circumstances of one's upbringing, certain personality traits will manifest themselves regardless.

The sharp division between these two groups, the *nurturists* (environmentalists or blank-slaters) and the *natalists* (who believe that biological inheritance is more important), has dogged developmental psychology for the better part of a century and it was perhaps inevitable that the debate should spill over into evolutionary psychology when this discipline emerged. Unfortunately, this most recent manifestation of the dispute has come to be characterized as a conflict between an evolutionary and a more conventional psychological approach perhaps, at least in part, because the environmentalist approach has been largely dominant within mainstream psychology since the 1970s. However, as we tried to make clear in Chapter 1, there is no explicitly *evolutionary* reason why we should prefer a natalist approach over a nurturist one. Whichever way evolution chose to produce the mind, that's the way evolution did it and it is an equally good evolutionary explanation either way.

separating the inseparable

In reality, neither genes nor environment can do the job on their own; rather, it is the interaction between these two influences that is important. This, the *interactionist view* developed within biology during the 1960s. While it is no doubt true that most people take the interactionist view on board and are careful to emphasize the importance of both genes and environment, it is also true that many sometimes fail to appreciate the full implications of this standpoint. This is reflected in the fact that, while accepting that both genes and environment contribute, there is often a desire – reflected nowhere more strongly than in the media – to partition behaviour into its genetic (or 'innate' or 'inherited') causes and its environmental causes.

This tendency probably owes its origins to misunderstanding the way behavioural geneticists present their findings and the measures they use. Behavioural geneticists are interested in the 'heritability' of particular traits, for example, athletic ability or sexual orientation. Heritability is a technical term, defined as the ratio between the proportions of variance in a trait that can be attributed to differences in genetic make-up or to differences in rearing conditions (the environment). To put it less formally, it is a measure of the extent to which it is possible to predict the distribution of a trait in the offspring generation, based solely on knowledge of its distribution in a parent population and the characteristic mating patterns of that population. To put it even less formally, the question being asked is: how well do the behavioural and physical traits shown by children correlate with those of their parents? If most of the variation in the trait can be predicted from the distribution of the trait in the parent population and the characteristics of the mating pattern alone, then the trait is said to have high heritability.

Height is one trait with relatively high heritability among humans: much of the variation in height can be explained by the height of parents (tall parents tend to have tall children). However, when behavioural geneticists make statements like this, they do *not* mean to imply that genes largely determine height and the environment plays little role. While behavioural geneticists are interested in how genetic variation influences the distribution of traits in a population, their methods make no assumptions or inferences about the manner in which a particular trait actually develops in an individual. Their focus is on the overall outcome of the developmental process, not on the developmental process itself.

The same is true of claims that a particular gene (that is, segment of DNA) is associated with a particular condition (schizophrenia, violence or a-grammatical speech). The fact that a particular gene is correlated with a particular condition does not mean it is the sole determinant of that condition. Only a handful of traits have such simple genetics as this (eye, hair and skin colour are some of these) and even then there are usually several genes involved. In most cases, traits are determined by a complex genetic cascade that involves a large number of genes as well as some aspects of the environment in which they develop, such as the order in which they are switched on and off. A gene may be involved in a particular trait not because it determines the trait but because it produces a particular effect that is crucial to the correct development of the trait. For example, if your car has a dodgy spark plug, it will prove impossible to start your car and drive around. Only the most naïve of us would, however, claim that the spark plugs make the car go. What is very obvious when talking about the behaviour of the average family car is, it seems, much less so when discussing our own behaviour and so we continue to make a very basic mistake that it wouldn't occur to us to make if we were talking about an inanimate object.

Newspaper reports which loudly proclaim that researchers have found the 'gay gene' or the 'language gene' are therefore making a fundamental error by assuming that heritability tells you the degree to which a trait is controlled by genes as opposed to the environment. It does nothing of the sort.

a cook's tour of interactionism

The point of the interactionist position can be seen more intuitively if you think about baking a cake. You assemble all the ingredients that are needed – eggs, butter, sugar, flour and the like – mix them together and place them in a hot oven. An hour later, if you're lucky, the wet sludge that you placed in the cake tin will have transformed into a light, fluffy sponge cake. So, is the fluffiness of your cake a result of its 'genes' (the ingredients) or the environment (the hot oven)?

What about its golden brown colour on top? How much of that can be attributed to what you put in it as opposed to what you did to it? Is your cake 80 per cent ingredients (genes) and 20 per cent oven temperature (environment)?

When thinking about cakes, these kinds of questions seem just plain daft. The cake is fluffy and golden brown because of the whole process: the interaction of the eggs, butter, flour and sugar with the mixing procedure and the time spent in the hot oven. You can't say how much of any particular trait that your cake possesses is due to a particular ingredient or how it was cooked. The heritability of fluffiness in cakes may be high, in that a lot of the variation in fluffiness can be attributed to differences in the type of flour used (equivalent to genetic differences) but that doesn't mean that flour alone determines the fluffiness of cakes. The same is true of people (and all other organisms). You were made by the interaction of genetic instructions with a particular set of environmental variables and, like a cake, you are greater than the sum of these parts and cannot be reduced to any of them in some simple-minded way.

The other implication of an interactionist view is that, if genes and environment interact to produce an individual, then, in order for a species-typical version of the individual to be produced, not only must the genetic information it inherits be similar to that of previous generations but the environment must be similar as well. This means that an evolutionary view that concerns itself only with genes is missing half the story: there must be environmental, as well as genetic, inheritance and continuity if evolution is to take place.

While Dawkins's 'gene's eye' view of the world has helped us to get to grips theoretically with some tough evolutionary problems, the downside is that the importance of environmental inheritance and the importance of the developmental process remain wholly implicit in these discussions. However, both developmental systems theory and, more recently, niche construction theory (in which the emphasis is on the extent to which organisms create their own environments and thus drive evolutionary change through a complex feedback loop) bring interactionism more firmly to the fore. An enriched understanding of behaviour, human and non-human alike, requires us to drop simplistic notions about the relative contributions of genes and environment and instead accept and appreciate that they form a complex whole that requires complex and sophisticated analyses.

The nature versus nurture issue has always been closely tied to the question of whether there are any 'human instincts'. Instincts are behaviours that are shared by all members of a species, which are innate, or genetically 'hard wired', and hence emerge without learning (often being present at birth) and which do not undergo any change once they have appeared. How does this square with the position we developed above – that it is not possible for behaviour to be wholly genetically determined? Don't instincts give the lie to this claim? If we do something by instinct, doesn't this mean that we are not in control of our behaviour?

This might well be true if instincts conformed utterly with the definition given above. A behaviour that appeared, fully formed, at birth (or at some point in early post-natal life) and underwent no change thereafter would indeed deserve to be called genetically determined. However, the truth of the matter is that most examples are not as clear cut as this and it is very difficult to establish whether a behaviour is unaffected by environmental influences during development and whether it remains impervious to change. While we may feel as though we never learned these behaviours and that they 'come naturally', eliminating all other possible influences is more difficult than it seems.

To illustrate the points we made above (and as a prelude to Chapter 4 where we deal with children's cognitive development in more detail), let us look at some examples of 'instincts' that are important during early development.

development and imprinting

Imprinting is a good example of an instinct that isn't all it seems. While imprinting can take many forms, the most well known example is the way in which young animals rapidly learn the details of their mother and form a social attachment to her. In essence, the young animal learns to follow the first moving object that it sees. Usually this is their mother but, as Konrad Lorenz famously showed, if the first moving object that young goslings see is not a mother goose but a human being, they will imprint on the human instead. One of the classic images from the history of ethology is that of Konrad Lorenz striding across muddy fields in his wellies leading a line of young geese dutifully following 'mother'.

Since this behaviour appears so early, it is often thought to be a hard-wired instinct, but in fact imprinting is an example of rapid, guided, learning. Young animals are primed to learn their mother's features, so that they can recognise her and not put themselves at risk by approaching another, perhaps less friendly, member of their own species. As each individual differs from another in unpredictable ways, it is not possible for the relevant information to be specified in a genetically-determined template. The youngster must learn the individually distinctive features of its mother and this is something that requires learning from experience. This is why chicks reared in abnormal circumstances can form social bonds to humans. It is the ability to learn quickly during a sensitive period that is genetically primed, not the ability to recognise 'mother', as such.

This rather nicely emphasizes the point we made earlier that, for evolutionary processes to work, environments, as well as genes, must be inherited. Imprinting works, under normal conditions, because chicks usually encounter their mothers as soon as they hatch and the interaction between the genetically-primed learning process and the normal developmental environment ensures that chicks form a protective social bond with their mothers (and, incidentally, vice versa!). If the wrong kind of environmental resources are present, then the chick will form an abnormal attachment, even though its genetic inheritance is no different from that of a chick that has imprinted on its mother.

As well as periods of rapid learning occurring shortly after birth or hatching, it is also possible for learning to occur before the event. This also makes the notion of instinct more difficult to define and identify: just because a behaviour is present at birth, one shouldn't assume that no learning has been involved. For example, immediately on hatching, young chicks show a preference for the maternal calls of their own species. This was taken to be an example of an 'instinct', in the sense of a behaviour that emerged without learning, since the chick hadn't had any opportunity to learn about the calls. However, in a classic series of experiments, Gilbert Gottleib (an early defender of developmental systems theory) showed that the chick's own vocalisations, which it makes whilst still in the egg, are very important for the development of this instinctive preference. If unhatched chicks are prevented from making calls and from hearing their own voice, then they are less able to pick out their own species' calls once they are hatched.

Closer to home, there is evidence that human babies learn to imprint on their mother's odour whilst in the womb. New-born babies respond very strongly to the smell of their own amniotic fluid: they cry much less when exposed to this odour, and they prefer to nurse at a breast that has been moistened with their amniotic fluid. This learning may help the baby to recognize and imprint on its mother after birth. Before widespread hospital births, with their clean and tidy procedures, birth would have involved mothers becoming covered in their own amniotic fluid, thus allowing the infant's preference to become associated with its own mother.

Babies can also recognise the smell of breast milk. Two week old babies that have never been exposed to breast milk prefer the smell of a gauze pad worn by a nursing mother to one worn by a non-nursing woman. It is not entirely clear to what the babies are responding. However, we do know for sure that milk has a specific odour, as do the areolar secretions of the mother's nipple. This preference for breast milk can be considered as an instinct, since it does not apparently involve any learning on the part of the infant. The fact that it is an instinct does not mean, however, that the behaviour cannot undergo change. By six days old, this general preference for breast milk has shifted to a specific preference for the baby's own mother's breast milk. Babies also prefer their mother's underarm odour to that of a stranger.

Even the mothers themselves are not immune to these processes. There is evidence, for example, to suggest that mothers begin to learn the particular scent of their baby whilst it is still in the womb. Each of us has a characteristic scent, which reflects our particular combination of genes. During pregnancy, a mother's signature odour becomes blended with that of her offspring (which differs from the mother's, since the baby has a different complement of genes) and she becomes increasingly familiar with this odour over the course of her pregnancy. This probably explains why mothers can recognise their own babies from smell alone, even soon after birth. Mothers who have spent as little as ten minutes with their new-born child can recognise their own baby's smell on a piece of clothing. Again, if you were to witness this, you'd probably be tempted to call it a human instinct, since it appears that a mother 'just knows' which baby is hers. But it is the learning process that the mother has undergone during pregnancy, and of which she herself is unaware, that is actually responsible.

the little bundle of instincts

Human babies show other behaviours that we would want to call instincts. In many cases, these are reflex responses to particular stimuli and often reflect the maturation of particular sensory systems. One of the earliest systems to mature is the vestibular system of the inner ear. This is the 'sixth' sense that enables us to perceive our bodily movements and our orientation in space, relative to gravity, and is therefore important for balance. The Moro reflex, in which a sudden change in position causes a young baby to extend its legs and fling its arms outwards with open hands, then slowly return to a flexed position, is first shown in the womb, around eight months into pregnancy (when the vestibular system first becomes functional). Since the baby is now able to detect its position relative to gravity, it can take action when it perceives its position in space changing suddenly (as, for example, when it is held an inch above the bed and dropped on to its back, during the standard midwife's test for the Moro reflex). A functional vestibular system also enables a baby to turn itself into the proper head-down position, ready for birth. Babies that have a defect in this system are more likely to be born feet first (breech birth), most likely because they can't tell up from down.

Stimulation of the vestibular system is good for a baby's development. In premature babies, it has been shown to stimulate growth and weight gain, as well as make babies behave less irritably, breathe more regularly and sleep more. Every parent knows that the way to get a baby off to sleep is to rock them or, if that doesn't work, take them for a spin round the block in the car. Motion stimulates the vestibular system and stops babies from showing 'disorganised behaviour' (flailing limbs, screwed up face, high-pitched crying) and puts them into a quiet, alert state. If the stimulation continues long enough, it decreases arousal altogether and the baby drops off to sleep.

Touch also has important developmental effects, allowing babies to thrive and grow more quickly than if they receive no such stimulation. The strong parental impulses to touch and explore their babies' bodies during the early months of life not only allows parents to bond emotionally with the baby but is also instrumental in promoting the baby's emotional and mental development. Rocking and cuddling babies can thus be explained on both proximate (why do we engage in a particular behaviour at a particular time?) and ultimate levels (what is the behaviour for? what is its evolutionary function?). Proximately, we might want to rock a baby gently in order to stop it crying. But rocking also serves the ultimate function of promoting the baby's survival, since it increases growth and weight gain, as well as putting the baby in an optimal state for learning about the world and developing its mental faculties.

motherese and social smiles

Another behaviour which appears instinctive is the way in which mothers (in particular) talk to babies. Despite most adults' aversion to 'baby talk', it is nevertheless the case that, confronted with a gurgling infant, whether their own or someone else's, most people begin speaking in a slow, high-pitched and highly intonated manner, repeating the same simple words over and over, often accompanying this with exaggerated facial expressions. This use of exaggerated and stereotyped vocal patterns has been observed in many different cultures across Europe, Asia and Africa and appears to be a universal parental behaviour. It is known as *motherese*.

While it may sound extremely daft, this 'sing-song' way of speaking is ideally suited to stimulating a young baby's hearing. The high intonation and use of simple words with repetition makes it easier for babies to distinguish the different parts of speech. The contrast between syllables is enhanced by the very large swings in pitch, whilst speaking slowly makes it easier for babies to process speech, since babies process auditory information at half the rate that adults do. Loudness also makes it easier for babies to distinguish a person's voice from the background of other noise, since their hearing is less sensitive than that of older children and adults. Finally, the high pitch falls into the frequency range to which babies are most sensitive from about three month of age. As silly as it sounds to us, motherese is the perfect way to grab a baby's attention and let it begin to learn the rudiments of language.

As well as these natural responses to the mechanistic properties of speech, babies in the womb form a preference for the pitch and intonation of their mother's voice. Anything that resembles the sounds the infant has heard in the womb will be good at getting its attention after it has been born. This initial preference is then reinforced by the fact that motherese is usually accompanied by other rewarding stimuli, such as positive facial expressions, physical contact and other forms of affectionate behaviour.

As with other parental behaviours, it is not entirely clear whether motherese is really instinctive or whether we learn this behaviour from witnessing others interacting with babies – or indeed, whether the babies themselves train us to speak in this particular way. Even the most curmudgeonly adult will eventually succumb to speaking in the way that babies like best. A lot of normal adult speech (especially that of deep-voiced men) causes distress in infants and, as soon as adults detect this, they begin to change their way of speaking, in an attempt to calm the infant. If they succeed in doing so, they will soon find themselves speaking in motherese. Thus, the feedback between infant and adults has mutually reinforcing results.

Similarly to touch and vestibular stimulation, there is evidence to suggest that motherese is good for babies' development. Marilee Monnot found that, in a sample of fifty-two normal, full-term infants, weight gain at three to four months of age was positively correlated with two measures of the intensity of the motherese used by their mothers. One was the prosodic component of the speech (its pitch, rhythm and musicality) and the other was the semantic content (the percentage of speech segments that were infant-centred), which may have reflected the intensity of attention that infants were receiving. Whilst the influence of other forms of stimulation can't be ruled out here, motherese clearly places babies in the same calm, alert state as do other forms of physical stimulation and, as a whole, this complex of parental behaviours is well designed to help babies grow and thrive.

Perhaps the most universal of all developmental milestones and the one that can truly be considered an instinct, is social smiling. In every culture, across the globe, babies first begin to smile at about two months, even if they are born blind and cannot see who they are smiling at (and indeed, have never seen a smile themselves). These smiles are proper social smiles, produced in response to particular stimuli. Before this, babies often smile spontaneously but these smiles are unrelated to any particular emotional state, being mouth-only smiles, caused by the spontaneous firing of neurons in the baby's brainstem. Indeed, babies smile like this in the womb and usually do so when they are asleep (perhaps because the motor neurons controlling facial movements are very close to the neural circuits that control sleep). By contrast, the social smiles that begin at two months of postnatal age are real smiles: they involve a specific muscle near the eye, the *orbicularis occuli*, which cannot be controlled voluntarily. When we are truly amused or pleased to see something, this muscle contracts and produces the crinkly-eyed smile that we all recognise as genuine.

So why does this ability appear at two months, rather than two weeks, or two years? As with the Moro reflex, the first appearance of this behaviour coincides with the maturation of a particular neural system, in this case, the myelination of the basal ganglia. Myelin is a fatty substance which coats the outside of nerve cells (and specifically the long projections of the cells, called axons, along which nerve impulses travel). This fatty coating insulates the nerve cell, in the same way that a plastic coating insulates electrical wires, enabling nerve impulses to be transmitted more quickly and with greater efficiency. Put very crudely, myelination brings a brain system 'on-line', enabling information to be transmitted between different brain areas and processed more effectively. In other animals, the basal ganglia are associated with the production of stereotyped social displays, such as courtship, dominance and greeting. It seems they are involved in a similar process in our own species, triggering one of our most important social displays.

As with motherese, the feedback between babies and their carers means that this instinctive smiling behaviour is soon modified by learning. When a parent receives a smile from its new offspring, this leads to an increase in parental attention as they attempt to induce their baby to produce another of these rewarding stimuli. Over time, babies learn how to modify their smiles according to the particular situation in which they find themselves, so that their range of smiles becomes elaborated and their facial expressions more varied. Blind babies, on the other hand, cannot capitalize on this essential feedback, despite their innate ability to smile. Their facial expressions become less responsive over time: they smile less and show less variation in the types of smiles and facial expressions that they produce.

the paradox of language

Like social smiling, language is a trait that bears many of the hallmarks of an instinct. Whilst there are some differences between languages, all children follow essentially identical schedules for learning to speak. First, they begin 'babbling' at around two months of age, repeating chains of sounds, starting with vowels (which sounds like cooing, as they emit long strings of *ooooh*s and *aaaah*s) and later moving on to consonants, so that by the age of ten months they are combining these into long repetitive strings (*mamamamama, nenenenene*). They then enter a stage where they make single word utterances, followed by two-word phrases, finally producing proper sentences from about four years onwards. Children acquire language swiftly and easily during this sensitive period but as they reach puberty the ability to acquire language with little or no effort diminishes and, by young adulthood, as we all know to our cost, language learning becomes a matter of hard work and effort.

Noam Chomsky, the world famous linguist and political theorist, was the first to point out that, despite their seeming variety, all the world's languages have the same fundamental structure – what he termed a 'universal grammar'. From this, he concluded that language was something intrinsic to the human brain. However, he did not take a particularly evolutionary view of this capacity and refused to see language as an adaptation in and of itself. Instead, he preferred to view language as simply an epiphenomenon (or accidental by-product) of having a large brain. More recently, the cognitive psychologist Steven Pinker has challenged this notion, arguing in his book *The Language Instinct* that language shows all the hallmarks one would expect of a trait produced by the process of natural selection: its good design makes it unlikely that it is merely a fortuitous product of our unusually large brain.

Like the other examples we have given here – and despite its instinctive nature – experience and environmental influences are vital for normal language development. Whilst we may possess a set of grammatical rules that is innately specified, the particular language that we acquire during childhood and the way we end up speaking it (our dialect or accent) are both functions of experience. If a child is isolated from normal social interaction (either as a result of abuse or by deafness), language does not develop normally and, if isolation is extreme, not at all. A certain minimum amount of experience with one's native tongue is needed so that the innate biases within the child's brain can be correctly tuned by experience.

The well documented and universal shift from learning individual words to full grammatical speech correlates with the maturation of the areas in the left hemisphere of the brain associated with speech. Wernicke's area, the part of the brain associated with semantics (the meaning of words and sentences) shows a peak in synaptic formation (connections between nerve cells) at 8–12 months, with myelination occurring around a year or so later. By contrast, Broca's area, the part associated with syntax (the rules by which language is structured, otherwise known as grammar) reaches its synaptic peak at 15–24 months, and doesn't myelinate until 4–6 years of age.

As well as these maturational schedules, children's brains are also extremely plastic: new synaptic connections are formed more easily and more quickly than in adult brains, enabling them to acquire new information at a rapid pace. It also means that children are less vulnerable to any kind of brain trauma than adults. Children below the age of about four can have their entire left hemisphere removed (the side that is genetically biased to show language function) and still learn to talk, read and write perfectly well, because their brains are plastic enough to allow all the necessary connections to be formed in the right hemisphere instead. But if the left hemisphere is lost at a later age, all language function is lost with it.

Babies show differences between the left and right hemispheres in response to speech sounds at around six months into pregnancy. We know this from recording the electrical activity of the brains (on EEGs, electro-encephalograms) of babies born prematurely. They can perceive speech better with their right ear (which, because of a quirk of ancient evolutionary history, connects to the left hemisphere) and musical tones better with their left ear (which connects to the right hemisphere). Whilst the left side of the brain is more dominant for language, the right hemisphere plays a central role in responding to the melodic qualities of speech (known as prosody) as well as to music. It seems that this differentiation is already present three months before babies are normally born.

Since babies can hear and process speech sounds in the womb, they can start to learn something about language before they are born. There is even evidence that babies in the womb can learn a preference for particular soap opera themes if their mothers sit and watch these programmes often enough. So, it should come as no surprise that they also show a preference for their mother's voice and are more responsive to a story that was read aloud by their mother during pregnancy than to a story that is completely new.

Babies as young as four days old also show a preference for their native language. In one study, French babies were shown to be more responsive to a female voice speaking in French than to the same voice speaking in Russian. When the recording of the voice was distorted so that the individual words could no longer be discriminated, the babies still showed a preference for the French-speaking voice – indicating that it is the rhythmic and melodic qualities of language, rather than the actual words, to which babies respond.

Babies are also able to categorize the individual sounds of speech (known as phonemes) in a manner similar to adults. Babies can easily discriminate *ba* from *pa*, for example. Indeed, they are able to recognize many more phonemes than adults. Japanese babies can tell the difference between 'r' and 'l' sounds, which adult Japanese people notoriously find extremely difficult. Babies, it seems, are naturally 'broadband', primed to distinguish a wide variety of speech sounds. However, as they get older, they lose this ability and can only distinguish the phonemes of their native tongue.

As with many other brain functions, it seems to be a case of 'use it or lose it'. At birth, the areas of 'neural' space that are tuned to distinguish particular sounds are more or less equal. With exposure to their native tongue, more of this neural 'space' is given over to distinguishing those phonemes that make up the language it hears, which, of course, are the very ones it will be important to be able to distinguish accurately later on, when the baby joins the social world.

Like smiling, language is an innate capacity that becomes 'tuned up' by experience. Engagement with other language speakers is essential for normal development but, thanks to their 'language instinct', children do not need to be actively taught to speak in the way that we teach them mathematics or geography. Their innate abilities guide them and, during the sensitive period of early childhood, they do it effortlessly.

While many of the things we do, even as adults, have the hallmarks of instincts (and we can, if we wish, describe them in these terms), our behaviour is not driven in a mindless way, by instincts. These well-rehearsed habits provide us with shortcuts that save cognitive processing time: we don't have to think about what to do, we just do it. This may be important when the environment is constant (or when we need to it be constant, as during a baby's early development) or life is at stake (as when we catch a glimpse of an out-of-control bus hurtling towards us) but every day is complex and unpredictable and we need to be able to find a balance between the conflicting demands of different components of the biological system. In the next chapter, we continue with our developmental theme and look in more detail at children's psychological development: specifically, at how children come to understand their social worlds.

summary

A full understanding of how evolution operates means appreciating that all organisms and their behaviours are produced by the inter- action of genes and environment. Not only is it wrong to try and separate the two, it is impossible to do so. This becomes apparent when we consider so-called 'instincts'; behaviours that appear to be fully formed and genetically 'hard wired'. Like any other form of behaviour, instincts are produced by the interaction of genes and environment. They involve learning and can be modified by experi- ence in the same way as any other kind of behaviour. They may appear early in development or only during certain sensitive periods and different instincts may appear at different times. However, it is hard to show that they develop without any environmental influence at all. We can call these behaviours 'instincts' if we wish to get across their automatic nature, and we can treat them as an essential part of what it means to be human but they are no more genetically deter- mined than are our preference for steak over salad or Loretta Lynn over Mozart.

how to make us human

In this chapter, we explore human cognitive development in evolutionary perspective, focusing on how children come to understand other people and their minds. Not only is this the most relevant area of cognition as regards our subject matter but it is also one which epitomizes the points made in the previous chapter: human children develop via a dialectical process between their inherent genetic endowment and the environment (culture) in which they grow up. A child abandoned alone on a desert island, like Robinson Crusoe, would not develop into a human person: a human mind cannot develop in isolation. To paraphrase the seventeenth century English poet John Donne, we are not islands but 'a piece of the [social] continent, a part of the main'.

how babies learn about the world

At birth, human babies are utterly helpless, lacking the ability to do even the simplest of tasks. The reason is not that babies are inherently stupid but that their brains and bodies are so underdeveloped. For most primates, birth marks the point at which the brain has reached a level of developmental maturity where the infant can get about on its own. Humans offer a striking contrast to this, because our babies are born about nine months earlier than we would expect on the basis of their brain size: in our case, unlike other primates (and most other mammals) a significant amount of brain growth occurs after birth. Human babies are born at an earlier stage of development, relative to other primates, in order to allow them to negotiate through the rather narrow female pelvis – something they would not be able to do if they stayed the course of gestation that any self-respecting primate would require for a brain of our size.

This is a problem we have inherited from our evolutionary past, when bipedalism narrowed the pelvis (to provide a more stable platform for the trunk during striding); since this happened several million years before the human brain began to expand so dramatically, our ancestors encountered an unexpected difficulty in having to squeeze a baby with a large head and sizeable body through a passage that was too small. The solution to this dilemma was to give birth earlier, to a less well developed baby, and complete brain growth outside the womb – something that obviously had significant consequences not only for the risks the baby faced, but also for the amount and intensity of parental care required.

This extreme helplessness means, among other things, that it is very difficult to test babies' cognitive abilities. However, rather than considering them merely inscrutable, many early philosophers and psychologists took their apparent inability to engage with the world as a sign that babies knew nothing. If this were really so, how could a baby ever begin to learn anything? How would babies know where to begin?

In the 1930s, one of the pioneers of cognitive developmental studies, the Swiss psychologist Jean Piaget, came to the conclusion that babies did not, in fact, face quite such a mammoth task as it seemed: they actually knew quite a bit about the world as soon as they arrived in it; their minds (and, by implication, their brains) were designed to perceive and respond to some things and not others, and this structure then guided them along an appropriate developmental pathway. Piaget argued that, far from being empty vessels waiting to have experiences poured into them, babies are primed to reach out to the world and actively engage with it. They use these abilities to build up their knowledge in stages, using the skills they acquire at one stage as a platform from which to build a more sophisticated knowledge at the next, culminating in a symbolically based and rational understanding of the world.

Although some of Piaget's ideas about children's development have been revised and revisited in the decades since his original studies were conducted, his suggestion that children's minds possess a structure from the very start and that children use this as a basis from which to construct their own knowledge has stood the test of time well. This confirmation is largely due to some inspired experimental studies of very young babies. The key that unlocked the baby's mind was the dummy (or pacifier). The more interested a baby is in a particular event or object, the harder it will suck on a dummy, reflecting an increase in its level of arousal. Similarly, the more interested a baby is in an event or object, the longer it will stare at it. Once researchers cottoned on to the reliability of these responses in infants, it became obvious how to design experiments that would allow babies to 'tell' the experimenters the answer to their questions.

The basic design of such experiments is what has come to be known as 'habituation–dishabituation': infants are repeatedly presented with a stimulus until it fails to produce an increase in sucking intensity (they 'habituate' to it, to use the formal term). They are then presented with a second stimulus. If the baby perceives this stimulus to be sufficiently different from the previous one, it will 'dishabituate' and show an increase in sucking intensity. If, on the other hand, the baby doesn't perceive any difference or notice anything unusual in the new stimulus, then there will be no dishabituation and consequently no increase in sucking.

Using this experimental design, developmental psychologists including Elizabeth Spelke and Karen Wynn have been able to show that very young babies understand physical concepts like gravity and the solidity and permanence of objects (that is, that two solid objects cannot occupy the same space at the same time). For example, after habituating to a video showing a ball falling from a surface and dropping to the ground, infants were then shown a video in which, after rolling of the surface, the ball remained suspended in mid-air. Infants dishabituated immediately to this stimulus and looked at this impossible event for much longer than the normal sequence, suggesting that they understand that balls should fall and are 'surprised' when they do not. Babies also show evidence of being able to perform simple arithmetic. When infants were shown a sequence of events in which two dolls were placed, one by one, behind a screen, they looked significantly longer when the screen was drawn back to reveal only one doll (an impossible event) than they did when (as expected) two dolls were revealed. In other words, they seemed to 'know' that $1+1 = 2$ and were surprised when the sum didn't 'add up'.

Experiments like these have led researchers to suggest that babies enter the world equipped with a basic knowledge of three key domains: physics, biology and psychology. Obviously, this isn't to suggest that babies know that they possess this knowledge, or that they can 'do' physics, merely that they respond different to certain stimuli, in a manner consistent with how the world works. With this basic knowledge in place, children's understanding is constrained to develop in a particular way – a way that has been selected by evolution to ensure they become adults who show adaptively relevant behaviour. Without the constraints of this basic structure, children would not know where to start to understand the world around them. Just as scientists develop theories to guide their observations and experiments, so children use their basic theories of the world to guide their learning. Science would never progress if scientists collected data at random and then tried to construct a coherent theory from a jumble of facts. Similarly, children could not progress without 'theory-formation' mechanisms to guide them towards learning things that are most likely to be relevant to their understanding.

In this chapter, we're concentrating on the psychological domain, not only for its relevance but also because it presents us with an ideal opportunity to show that many of our psychological traits have a very ancient history. Despite the key relevance of culture and language to human development, the primary adaptation that kicks the whole thing off is a biological trait that has clear links to our primate cousins.

This is a point worth emphasising, since, as we discussed in Chapter 2, one of the assumptions that some evolutionary psychologists work under is that the psychological mechanisms we show today were selected for in what is known as the 'environment of evolutionary adaptedness'. Our evolutionary history extends much further back than the last few million years or so and many of the psychological adaptations that we possess have their precursors among our ape and monkey cousins. By viewing human psychological development in a comparative framework, we can see how many of our most distinctive traits can be traced to these ancient precursors and how the interaction of these traits within the human cultural milieu results in our particular brand of social cognition.

the eyes have it

Within minutes of birth, babies are more likely to follow face-like stimuli than scrambled or random patterns. Tests performed on two month old babies have also shown that their attention tended to be focused for longer on face-like stimuli which showed the eyes compared to stimuli that did not. Like the studies investigating babies' understanding of the physical domain, these studies reveal that even very young children have a knowledge of the world they have so recently entered; in this case, a basic understanding of other people. Some researchers, for example Simon Baron-Cohen of Cambridge University, suggest that the ability to detect and focus on eyes is something that is hardwired into the brain and present at birth, similarly to a reflex action.

However, other studies have shown that babies actually prefer to look at any visual stimulus that has up-down asymmetry (where the majority of pictoral elements are concentrated at the top, rather than the bottom). In this view, babies do not have any hardwired preference for eyes or faces as such but for any stimulus that shows a top-heavy pattern and this guides them into attending to faces because faces (usually the mother's) are one of the commonest things of this kind they see at close quarters. In fact, babies' abilities to recognize and interpret facial expressions seem to be quite poor when they are very young, which suggests that learning is in fact needed to distinguish some of the finer points about faces.

Neurobiological studies of monkeys suggest that the populations of neuronal cells active in object recognition can be progressively tuned up in response to exposure to particular stimuli, becoming increasingly more specific to the stimuli in question (that is, less likely to respond to other, similar, stimuli). It seems that, like monkeys, babies have areas of the brain that gradually develop a specialized response to faces and eye gaze, enabling more precise recognition of expression and gaze direction as they get older.

This focus on faces is made possible by two specialized aspects of primate neuroanatomy. Studies of rhesus macaques monkeys, by David Perrett at the University of St Andrews, have revealed that an area of the brain known as the anterior superior temporal sulcus contains neurons that show highly specific responses to particular kinds of social stimuli. For example, there are cells that respond only to a head facing left, not to one in any other orientation. There are also cells that respond purely to faces, facial expression and eye gaze direction (in particular, to eye gaze directed towards the subject). There are also cells that respond exclusively to biological motion (that produced by a living being rather than by an inanimate object).

The second important feature is that the primate visual system has two distinct pathways along which which visual data are transmitted from the visual areas at the rear of the brain to the frontal lobes where the information is processed further. The *magno-cellular* system is concerned with movement detection and is common to all mammals but the *parvocellular* system is unique to primates and analyses fine detail and colour. The magnocellular pathway passes over the dorsal (top) area of the brain, while the parvocellular pathway follows a ventral (bottom) route and is linked to the amygdalae (a pair of almond shaped structures buried deep in the temporal lobes) which are involved in the perception and processing of emotions. Robert Barton of Durham University has shown that, amongst primate species that are active only during daylight, the number of cells in the parvocellular layer is positively related to social group size but the number in the magnocellular layer is not. Barton suggests that the parvocellular layers were enhanced during primate evolution in order to process the fine details of dynamic social stimuli, like facial expression, gaze direction and posture. The connections through the amygdalae are especially significant, because they allow an emotional 'signature' to be attached to the signal.

Human children, like their primate relatives, thus seem evolved to respond to socially relevant visual signals. However, as infants develop, a crucially important difference develops in the way they respond to these socially relevant stimuli, compared to their fellow primates. Between 9–14 months of age, the tendency for a young human infant to focus attention preferentially on the face-like stimuli in its environment develops into an ability to follow the gaze of adults and to look where they are looking. At this point, they begin to show a phenomenon known as *joint attention* whereby they use another individual's gaze direction to focus their own attention on the same object. They also begin to engage in *shared joint attention*, where they look from the object to the other individual and back again, to check that they are both attending to the same object. In addition, by 14 months of age, children are capable of directing an adult's attention to the object they are looking at (for example by pointing), so that the adult's attention is co-ordinated with theirs. This shows that not only can human infants tune into the attention of others but that they also know how to get adults to tune into them.

While this may seem like small potatoes, no other primate seems capable of doing this: monkeys and apes in the wild have never been seen to point at an object purely to draw another's attention to it. Nor do mothers and offspring show any evidence of sharing attention. Although they care for them, groom them and carry them around constantly, monkey and ape mothers spend hardly any time actually looking at their babies' faces. This is in stark contrast to most human mothers and infants, who happily spend hours gazing into each other's eyes, having 'conversations' with each other (often just through the exchange of facial expressions).

It is clearly no coincidence that nine months is frequently the point at which parents start commenting on how their baby is developing a personality. What parents are picking up on here is their infant's increased responsiveness to them and its ability to connect with them through shared attention. This makes the baby feel more like a person than an eating and (not) sleeping machine.

the cultural ratchet

So the ability to engage in shared attention seems to be a distinctively human characteristic. Michael Tomasello, a leading developmental psychologist, whose theory of development will be our focus here, goes so far as to argue that shared attention is the key cognitive skill that distinguishes humans from the rest of the animal kingdom. Sharing attention, Tomasello argues, shows that, from nine months of age, human infants understand other humans as *intentional agents* (that is, having intentions to act in particular ways) 'whose perspective on the world can be followed into, directed and shared'. In other words, human babies are able to recognize and respond to the psychological state of others. Tomasello sees this as the key human cognitive trait because from this, all else follows. In order to produce a competent human person, all that is needed is for children to develop the basic skill of shared attention, which allows them to connect to other humans and thus become immersed in a human cultural world. He calls it the 'nine month revolution'. From this point, the interaction between the child and its cultural environment leads to the flowering of the cognitive skills that make a fully-fledged human, including language, imitation, empathy and co-operation.

How does this happen? Very simply, once children are sharing attention, a feedback loop is set up, whereby the more responsive the infant becomes and the greater feedback it gives the adult, the more time and effort adults will spend in engaging with the baby, thus increasing the amount of stimulation it receives and further improving its ability to respond. Tomasello calls this the 'rachet effect': as children acquire one set of skills, this changes the way in which adults interact with them, such that children can then be led into a more sophisticated form of interaction, which in turn allows them to acquire further skills. In this respect, Tomasello's work builds on the work of Lev Vygotsky, a Russian developmental psychologist, who first pointed out the importance of the social world to children's cognitive development but it also owes something to Piaget's view that children construct new knowledge on the basis of knowledge that they acquired earlier in the developmental period.

Tomasello views the capacity for shared attention as a specific biological adaptation that humans, but no other primates, have evolved. He argues that this is essential for the cognitive ratchet to move on. However, the philosopher Matteo Mameli argues that it is also possible that shared attention in babies is the result of the ratchet operating at an earlier stage than Tomasello suggests and may be a consequence of what he terms 'expectancy effects' on the part of the adults who interact with them. Adult humans have very good mind reading skills: we readily infer and attribute mental states (like beliefs and desires) to other individuals, to explain their behaviour. Mind reading of this kind leads to what Mameli terms 'mind shaping', where the beliefs you hold about another person lead that person to act in a way that fits your beliefs about them. For example, if you believe your friend, John, is angry with you, you may find yourself behaving towards him in a way that is rather hostile and defensive. This hostility on your part may lead John to feel angrily towards you, even though he wasn't in the least angry with you to begin with. Another term for expectancy effects, one that may be more familiar, is that they are 'self-fulfilling prophesies'.

Mameli's position is that children may not really be engaging in shared attention initially but merely displaying a simpler reflex response to faces and eyes. This action on the part of the baby leads adults to believe, hope and expect that the baby wants to communicate with them and so the adult behaves toward the child in such a way (holding gaze, making faces) that the child is able to learn that eyes and faces can provide them with information about the world, resulting in the child learning how to share attention. The adults' belief that the baby wants to communicate is a self-fulfilling prophecy; as a result of this belief (whether it is accurate or not), they behave in a way that enables the child to acquire the ability to share attention and therefore to communicate. Tomasello sees the ratchet begin to move on once children have acquired shared attention; Mameli sees it starting before then, through the actions of the adults with whom the baby interacts.

Mameli describes this as an example of niche construction; by interacting with the environment, organisms change it and so can modify the selection pressures acting on themselves and their descendants. An important part of the human niche is other people and, via mind shaping expectancy effects, we modify the niche of the humans we interact with. Other minds are thus a reliably recurring developmental resource that human children need in order to become human adults. The belief that children possess an ability to communicate socially, rather than any actual ability on their part, may play a large part in causing Tomasello's ratchet to move on.

The key question here is: how does this all get started? If mind reading skills in children cannot develop unless there are mind shaping adults already in place, how could mind reading have evolved in the first place? Mameli's argument is that only a very basic ability to read minds is needed. All that is required is that parents see something in their child's behaviour that leads them to believe the child is trying to communicate and to which they are responsive; the self-fulfilling prophecy does the rest. If this difference in developmental experience produces better psychological skills in these offspring, then they will be even more likely to believe the same things about their own offspring, perhaps generating even greater expectancy effects, and influencing their children's development more strongly. The evolution of mind reading and mind shaping skills could then proceed as a classic example of niche construction. The niche constructing trait of mind shaping would co-evolve with traits that respond to mind shaping, such as joint attention, resulting in a mutually reinforcing feedback process that leads to an evolutionary spiral towards the kinds of mind reading abilities that all humans share and which are now an essential component of normal human development.

mother knows best

Once children are capable of joint attention, they begin to use *social referencing* to guide their behaviour. This is the means by which young babies gauge the feelings of another individual (usually the mother) towards an object in the environment and then use that information to form their own attitude toward the same object. When they are unsure about something (a new toy, a strange person), infants glance rapidly back and forth between the object and the mother's face, in order to read her expression. If the mother shows any sign of uncertainty or fear, the baby will not attempt to play with the object. If, on the other hand, the mother smiles and encourages the baby, then it will happily begin to explore the interesting new toy or approach the strange person. The infant's behaviour is utterly dependent on the mother's expression.

Another behaviour that begins to make its appearance in the first year or so of life is *imitative learning*. Unlike other monkeys and apes, human infants can apparently recognize the difference between the goal of a task and the means used to obtain it. That is, they can understand why adults will use different behaviours to achieve the same goal (because, for example, the usual way is blocked) and do not see the means used as intrinsically linked to the goal. In a similar vein, 16 month old infants are more likely to imitate intentional than accidental actions, recognizing the difference between a goal-oriented action and one that serves no useful purpose. By the age of 18 months, when they see an adult trying to do something, children will reproduce what the adult was trying to do and not what they actually did, such that the infant succeeds in the task that it watched an adult fail.

As with joint attention and social referencing, imitative learning shows 'the mark of the mental' as Tomasello puts it, revealing that children understand others as intentional agents and recognize that the goal-directed actions of others are equivalent to their own actions under the same circumstances (they show *self-other equivalence*). Human children are thus capable of *true imitation*; they can understand the intentions behind a sequence of actions and replicate them precisely.

As we mentioned in Chapter 2, this ability is one which earlier human species did not apparently possess; the hand axes of *Homo erectus* seem to have been produced by *emulation learning*, where individuals achieved the same goal, of producing an axe but did so in an idiosyncratic manner, preventing true cultural learning via the same kind of ratchet effect that drives social development. Tomasello argues that the advent of true imitative learning is what leads to cultural behaviour as we know it: through recognizing both the goal and the action sequence used as relevant to a task, children come to understand what an object is 'for' in their particular culture. They begin to recognize that functions are assigned to objects by collective agreement, not randomly or at whim.

let's pretend

This becomes even clearer to children as they begin to engage in pretend play. From around 24 months of age, they are able to use a banana as though it were a telephone or hold dolls' tea-parties, where, if they accidentally spill 'tea' on one of the dolls, they dry her off because she's 'wet'. Young children can also recognize and understand when other people are pretending and will happily join in with their games. As they engage in these sorts of activities with adults, children gradually learn that a change in their perspective and that of others can modify the function of an object ('this banana only counts as a telephone while we're playing'). This, according to Tomasello, is an important cultural process, because learning to use toys for pretend play is an example of a 'socially constitutive process' where the adults and children dynamically create the function of an object by collective agreement. We'll return to this in the chapters dealing with culture, where the notion of socially constituted, as opposed to objective, facts can be seen as the key defining feature of human culture.

So, by the age of two, children can understand the attention and intentions of others around them and use this understanding to gain knowledge of the world. This process is helped along by language, which, as we mentioned in Chapter 3, children start to acquire in earnest from about 18 months. Again, while language is a complex cognitive function, it stems from the ability of children to engage in shared attention. When they first begin to learn names for things, children associate the sounds they hear with the object that they and the adult are looking at. More importantly, they also begin to understand that words are 'bi-directional' or socially shared: once children understand that an adult, by making a particular sound, is attempting to direct their attention to a particular object, they also realise that, if they want the adult to share attention to the same object, then they can use that sound for the same purpose.

What is also striking is that two year olds are beginning to appreciate that others can have a different perspective on something to the one they themselves have. They can correct themselves and use a more appropriate word for an object when they realize that the other individual won't necessarily know to what they're referring (for example, 'the man... the policeman'). All this ability stems from the 'nine month revolution' and leads on to what is, from an evolutionary point of view, perhaps the single most important cognitive revolution of childhood.

reading minds

By the age of four, children not only understand that people are intentional agents, but also begin to grasp that they are *mental agents* as well. From understanding simple mental states like attention and intention, they now begin to understand more complex ones, like beliefs and desires. Thus, they realize that people's behaviour is driven by what they believe the world to be like and that these beliefs do not always accord with reality. This is often known (perhaps a bit misleadingly) as *theory of mind*.

Philosophers of mind use the term *intentionality* to refer to the cognitive processes involved in this ability. Intentions, in this philosophical sense, cover mental states characterized by terms like *believes, intends, supposes, infers, wants*, etc. Intentionality, or the *intentional stance*, refers to a reflexively hierarchical series of mind states. Organisms that know their own minds ('*I believe that ...*') are said to have first order intentionality; having a belief about someone else's mind ('*I believe that you suppose that ...*') is second order intentionality; third order is having a belief about someone else's belief about one's own mind ('*I believe that you suppose that I intend that ...*'). Theory of mind corresponds to second order intentionality.

The classic test used by developmental psychologists to detect whether children have made the transition to theory of mind (understanding people as mental agents) is the *false belief task*. This is deemed to be a benchmark of theory of mind because it can only be answered correctly if the child can differentiate its own knowledge of the world from the beliefs held by another – beliefs that the child must assume, on the basis of its own knowledge, to be false. It must have a belief about someone else's belief.

The classic false-belief task is known as the Sally-Ann test. Sally and Ann are two dolls manipulated by the experimenter. Sally places a ball in a basket and then leaves the room. Whilst Sally is away, Ann takes the ball from the basket and hides it in a box. The child is then asked: 'When Sally comes back, where will she look for her ball?' Children under the age of four fail: they typically say that Sally will look in the box. They are unable to take Sally's perspective and understand that her mental state corresponds to a different version of reality than their own. Children over the age of four, however, almost always get the answer right and state that Sally will look in the basket, where she thinks the ball is, rather than in the box, where they know it is hidden.

Some developmental psychologists, like Josef Perner and Janet Astington, believe that this shift in understanding represents the key revolution in children's cognitive development. Perner, in particular, argues that there is a major reorganization of the child's knowledge at this age, the result of which is that children come to understand mental representation as a process. While two year olds may understand that representations are mental entities ('thoughts'), they do not understand that representation is also the process by which these mental entities are formed in a person's mind; they understand that the mind contains thoughts but they do not know how they get there. As a result, two year olds do not understand misrepresentation – that someone's beliefs can be false – because they do not realise that a person constructs the world in their head, rather than seeing it as it really is. Hence, two year olds can understand that Bob has a representation *of* his cake but they cannot understand that he represents the cake *as* being in the fridge (that is, that he forms a belief about the location of the cake). Four to five year olds, on the other hand, have no trouble with these kinds of problems.

Tomasello disagrees, he suggests that all the really hard work has been done during the first two years of life and that by this point children know lots of important things about what other people see, do, intend and attend to. Tomasello suggests that, rather than a representational revolution suddenly occurring at four, children's understanding of complex mental states is gradually acquired over several years of continuous interactions, particularly language-based interactions, with adults, following the 'nine month revolution'.

He suggests that what four year olds are coming to terms with at this stage is not that other people have beliefs that may differ from their own – children seem to have some grasp of this at earlier ages, as revealed by their ability to use language appropriate to another's perspective – but with the notion of *reality itself*. That is, they come to understand that there is an objective reality which exists independently of their own and others' beliefs. Reality is not their personal perspective of the moment (which may conflict with another person's) and it is not the perspective they may happen to share with another person but is a 'view from nowhere' – one that doesn't depend on anyone's point of view.

Tomasello believes that children can't understand the notion of false belief properly until they understand how objective reality, subjective beliefs and the beliefs shared between themselves and others are integrated. Once children understand this, as four year olds appear to, then they appreciate how one person's views may differ not only from their own view but also from the way the world truly is.

He also suggests that four year olds begin to understand that certain facts about the world are not only created during the social interactions they share with another person (as when they engage in pretend play with an adult) but also by the wider culture in which they live, through shared beliefs and practices; in so doing, they move from what Tomasello calls 'shared intentionality' to 'collective intentionality'. Two year olds cannot, for example, grasp concepts like marriage or money, since their viewpoint on the world is still too narrow, restricted, as it is, to their point of view and those that they share, moment-by-moment, with another. Four and five year olds, by contrast, have broadened their horizons, through their new grasp on reality; they realize that there are viewpoints which extend beyond their own, very personal, domain of interaction and which are shared by all people in their culture. It is at this point that children begin to lose the rampant egocentrism that has characterized their lives until this point; it stops being all about 'me' and starts being all about 'we'.

By six to eight years of age, children have acquired much of what they need to navigate through their social and cultural worlds. Of course, development continues; indeed, it is lifelong – we are never too old to learn something new. But there is some truth in the old Jesuit saying with which we began Chapter 3: the first half-dozen or so years of life represent the time when we acquire many of the critical skills we need to participate in, and create, human culture. Miss out and we can probably never be fully functioning members of our community.

The key cognitive skill of shared attention, which has its precursors in the visual cognition of our primate cousins, sets the stage for the entrances and exits we will make over the course of our lifetime and is in place at only nine months old. From this point, it becomes impossible to separate biological from social development. Without the ratchet effect provided by the interaction between a child and its culture, it would not be possible to make a human mind. Our essential nature is contained in the way that we are nurtured and cannot be divorced from it. A cake is only a cake because you cook it.

summary

Children's ability to understand the social world they inhabit requires the interaction of evolutionarily ancient adaptations inherited from our primate ancestors, the mind shaping influences of other human beings and the culture in which they live. Human babies are born with basic knowledge of the world and this knowledge is built upon and shaped by the actions of the babies and those around them. By the age of nine months, human infants have acquired the ability to engage socially with others through joint attention, an ability not seen in other primates. With this in place, children's social understanding increases via a 'ratchet effect', where the social knowledge they gain at each stage provides a platform for further improving and increasing their knowledge of others. From joint attention, children move on to imitation and pretend play. Finally, by five years of age, children are able mind read for themselves, understanding that beliefs about the world and the reality of the world are two very different things.

choosing mates

In the previous two chapters, we considered the cognitive abilities of children and how they develop. However, children's development is the end of a much longer process that involves many complex decisions by parents, one of which is how many children they should have and how much love and care they should lavish on each. But even before they get to this stage, the parents must agree to choose each other as mates. It's probably fair to say that these are the two biggest decisions that we make in our lives, and we now turn to consider them in some detail. Although we'll discuss parental investment decisions in more detail in the next chapter, it will be hard, when discussing mate choice, to avoid the topic entirely, because willingness to invest in offspring is one criterion we use in choosing our mates.

Mate choice decisions are not simply a case of casting one's eye around the room, choosing a member of the opposite sex at random, then setting up home and having children. The process of mate choice is one of negotiation and the decisions that people make depend on a number of factors. For one thing, even if it were just about finding the one person to whom you would wish to declare your eternal devotion, there is no guarantee that he or she will feel the same way about you. Most people have particular likes and dislikes and the attributes that you have to offer might not be quite what they had in mind.

But, before we get into the specifics of mate choice, it's important to set out their constraints and the context within which they play themselves out.

the constraints of ancient biology

For better or worse, the fact that we are mammals imposes an asymmetry in the costs of reproduction between the two sexes. This asymmetry has profound consequences for male and female sexual strategies. Some time before the dinosaurs became extinct (around 65 MYA) and a very long time before the first hominids stepped gingerly on to the African plains, our mammalian ancestors set us on a path from which we have not been able to deviate, when females opted for internal gestation and lactation. This has one crucial consequence for male mammals: they can contribute little more than sperm to the business of reproduction. Females, on the other hand, first gestate the foetus and then, in the form of milk, provide it with the nutrients and energy required to fuel its postnatal growth.

The contrast between the two sexes' involvement in reproduction necessarily has implications for how each sex views the business of mate choice. Once a female is pregnant, she can't get pregnant again, no matter how many males she mates with. Additional matings do not increase the number of offspring she can carry and may even be detrimental to her and her offspring's health. For males the picture is rather different. Because male reproductive investment is cheap, consisting mainly of the time taken to secure a copulation and the contribution of easily replaceable sperm, males can increase their fitness by attempting to mate with, and impregnate, as many females as possible. All else being equal, females will tend to emphasize rearing, to make sure that all goes well during this long period of investment, whereas males will tend to emphasize mating opportunities.

This sex asymmetry has several important consequences. First, females will have more to lose by making a mistake in mate choice, so they ought to be more choosy than males. If a male makes a bad choice, all he's lost is some time and a drop of sperm; females, by contrast, are literally left holding the baby and, in the case of human females, who have a finite number of eggs and a reproductive life span that is terminated by menopause, each pregnancy represents a significant fraction of their reproductive life span. Second, around 90 per cent of mammal species are polygamous, that is, one male mates with many females and in many cases devotes a great deal of time and energy to keeping rivals away from his mates. Monogamy, by contrast, is rare (except in the dog family, where it is the rule).

In humans, the situation is not quite so straightforward. Human mating systems involve pair bonds which, if not entirely monogamous, do tend to encompass a more co-operative engagement in the business of rearing. This reflects the demands of the slow development and long period of dependency of human offspring (itself a consequence of their large brains). Since human males are somewhat constrained in their ability to seek additional matings with other females, it is perhaps not surprising to find they emphasize fertility as a criterion of mate choice.

the rose-tinted world

So, what, from a purely biological point of view, should a woman look for in a man? Bearing in mind their high reproduction costs, they should preferentially mate with men on the basis of their effect on the success with which children can be reared. This reduces to two main considerations. One is the quality of their genes: better quality genes usually translate into better quality, or more successful, offspring. (Evolutionary biologists refer to this as choice for 'good genes', meaning genes that beneficially affect the biological quality of offspring, without meaning to imply any moral or social value judgement.) The other is men's ability and willingness to contribute to childcare, either directly, by contributing to the child's development or indirectly, by providing for the mother and child. How women balance these two considerations depends entirely on how strongly each affects their ability to rear offspring successfully.

If female mate choice involves an evaluation of males on the basis of their genetic quality, then we should expect to find predictable cues to male quality. Such indicators seem to include general body symmetry, symmetrical facial features and scent (criteria that are also known to be important for mate choice in birds and other mammals).

If, on the other hand, females choose mates on the basis of their ability to contribute to child-rearing, then this can take one of two forms. To the extent that successful rearing of offspring ultimately depends on provisioning, the resources that a male has to offer, now or in the future, will be important considerations. In traditional hunter-gatherer societies, those resources might translate into the man's hunting abilities but in agricultural and industrial societies they are more likely to reflect his intrinsic wealth and/or social status. We know only too well, from contemporary industrialized societies, that family wealth directly affects children's morbidity and death rates, and there is even more striking evidence for this in traditional and pre-modern agricultural societies where the best predictor of infant survival is the family's landholding (usually inherited through the husband).

Rearing investment is not, however, limited to indirect investment. There is the whole long business of socialisation, of teaching children how to be effective members of the community and of placing them in the most advantageous position in society that one can manage. Hence, a man with little to offer on the wealth dimension might still be an attractive proposition if he showed qualities associated with being a good parent.

Not only should women be looking for honest cues of these qualities but they should also be interested in a man's motives and commitment. Flashing a few diamond necklaces about now doesn't guarantee that he'll still be doing so in ten or twenty years' time. Given that human development is *so* slow and that the period of parental investment extends well into the child's second or even third decade, cues that signal long-term commitment may be as important as cues that merely signal the ability to invest. Such cues are likely to involve evidence of the male's willingness to invest in the relationship and might also include features such as social skills (being considerate, having a sense of humour, being a good listener) that will be needed to keep the relationship alive.

If those are the characteristics that women are expected to look for in a man, what should a man look for in a woman? Since monogamous pair-bonds (which characterize most human mating arrangements, even in the 80 per cent of cultures which permit, or approve of, polygynous marriages) limit males to one female at a time, males should be sensitive to indicators of female fertility. Since they are (theoretically) going to be spending a long time with a single female and hence forgoing other mating opportunities, they will do best if the woman they choose is the most fertile they can find. Men should thus express a preference for women who possess those traits that correlate with fertility, the most obvious (and highly correlated) of which are age and beauty.

The reproductive period of human females is much shorter than that of males, since it terminates at the menopause. Under these circumstances, if males choose females mainly on the basis of their fertility, then younger should be better than older. By the same token, when males marry polygamously, they should replace the current wife with a 'younger model', which should result in men progressively targeting women who are younger than themselves as they age. This will be true even in modern Western societies, where monogamy is legally enforced: in these cases, divorce and remarriage allow men (and women) to become serial polygynists and, in successive relationships, men (but not necessarily women) should target women who are progressively younger than themselves.

If these are the qualities that men and women should look for in a partner, how do we go about finding the ideal mate? There are a number of ways to examine mate-choice decisions in the real world and we'll consider just two. However, first we need to draw an important distinction between mate choice *preferences* and mate choice *decisions*. The distinction hinges on the fact that both parties to the arrangement have preferences: the fact that I like you does not even remotely guarantee that you will like me – I may not have enough to offer you, no matter how close you are to my ideal. Since we live in an imperfect world, we have to make compromises. Consequently, the qualities of the person we finally settle for will always be a compromise between preference and opportunity. You may settle for me, even though I am less than ideal, because there simply aren't any better alternatives within a reasonable distance. Or you may settle for me because, on balance, I offer the best combination of qualities – not too good on the wealth side, but not so bad on cues of genetic quality (such as physical attractiveness) or good on cues for commitment.

opening the bidding

One of the most successful ways that evolutionary psychologists have examined mate choice preferences is through analysing lonely hearts advertisements. These advertisements contain information about the individual advertiser as well as information about the kind of partner they are seeking. Unlike marriage data, which represent the final decisions that people make, lonely hearts advertisements provide a opportunity to uncover idealized preferences. In this respect, they have been described as 'opening bids' in the mating poker game, or as 'mate choice in the raw'. The advertiser is saying: 'This is my ideal partner, this is what I'd like'.

Analyses of advertisements from around the world have reported consistent patterns in mate choice preferences and these conform very neatly to those predicted from evolutionary first principles. In general, women who place advertisements emphasize two kinds of cues that they seek in men: they tend to express preferences for males who have wealth or status and those who are willing to invest in the relationship. By contrast, advertisements placed by men are dominated by preferences for female physical attractiveness; a reliable index of fertility in women (at least in traditional societies that lack the advantages of cosmetics and cosmetic surgery).

Importantly, there is a close symmetry between male and female advertisements. The qualities that people seek in a partner are mirrored in the self-descriptions of the advertisements placed by the opposite sex (that is, the target audience). So, for example, while males express a preference for females who have particular physical characteristics, female self-descriptions tend to emphasize those very characteristics. Men's, conversely, tend to advertise the traits that women most often ask for.

Another clear and consistent pattern to emerge from lonely hearts advertisements relates to the preferred ages of prospective partners. Female advertisers typically seek men who are 2–5 years older than they are, whereas as men age, they typically seek women who are increasingly younger than they. In Western societies, the preferred age seems to be women in their late 20s – which happens to be the average age at which women in these societies reproduce for the first time.

real life's the best of a bad job

As we previously pointed out, what we want isn't necessarily the same as what we get. There is no guarantee that the attributes that we possess (and advertise) are going to be those that are sought by our ideal partner. And this leads to an important point in evolutionary analyses of behaviour: decisions about how people behave are always dependent on circumstances. In mate choice decisions, our ability to attract an ideal partner will depend, at least partly, on our individual standing in the mate choice market. The harsh economic law of supply and demand applies just as much to the mating market as it does to selling soap powder.

Consider the case of a young, attractive and healthy female. Since youth and physical attractiveness correlate with fertility and fertility is what men seek, then it follows that our young and attractive female will have more bargaining power than other women who do not rate as highly on those particular attributes. As she is in more demand than other females, she can afford to be choosier. Similarly, men who have more wealth or status to offer (one of the traits in demand by women) can afford to be more demanding and choosy than men who do not.

And this is exactly what happens. Using lonely hearts advertisements, Bogus Pawlowski and Robin Dunbar have shown that both males and females are sensitive to their respective 'market values' and adjust their demands of what they seek in a partner accordingly. For both sexes, they found a strong positive correlation between market value (expressed as the ratio of demand over supply) and number of traits demanded of a potential partner. Tribal societies where men pay a bride price to their intended's family also offer clear examples of the influence of relative market value. Variability in bride price payments can confidently be interpreted as a reflection of the perceived value of the bride and payments are significantly higher for younger and healthier (and hence presumably more fertile) women. Incidentally, in many societies, failure to produce children is one of the principal (and legally sanctioned) grounds for divorce and a refund of the bride price. It is also one of the main predictors of divorce in Western societies.

Analyses of actual mate choice decisions bear out what has been found for mate choice preferences. Almost universally, spousal age differences mirror exactly the patterns reported from mate choice preference studies and have done so for a long time. Studies of marriage data from historical populations around the world reveal the same pattern: spousal age difference typically increases as husband's age at marriage increases. There is also a clear tendency for females to prefer males whose status is better than their own, a tendency that leads to *hypergyny* (women marrying up the social scale). In historical populations, like the nineteenth century farming community in the Krummhörn on Germany's North Sea coast, women married much earlier if they were able to marry into a higher socio-economic class than if they married into their own class. The latter women married later because they preferred to wait as long as possible in case a better catch came along but eventually had to settle for the best they could do lest they end up being 'left on the shelf'.

Jane Austen's acute literary observations of the English rural county class tell exactly the same story.

honest cues

In lonely hearts advertisements, advertisers are free to say what they like about themselves. There is a common perception that advertisers are inveterate liars. Careful analyses of advertisements show this not to be true. Indeed, we would not expect it to be true, since the purpose of advertising is to meet prospective partners and form relationships. Advertisers who do not match their descriptions do themselves no favours. The real problem, perhaps, is that those answering advertisements over-interpret the descriptions they read and are then disappointed.

This is not to say that advertisers don't bend the rules. The name of the game, in this early stage of courtship, is simply to stay in the market and not be rejected too soon. Writing an advertisement that results in no replies is, as dating agencies constantly point out to their clients, rather pointless. Hence, it should come as no surprise that advertisers seek to stack the odds in their favour. This can involve sins of omission: women over 35 often don't declare their age, because doing so results in a marked decline in the number of replies they get. Suppressing information on their age allows older women to be more demanding of prospective partners, in the knowledge that there is a fair chance that, once a prospective partner meets them, they will trade off their age against their other qualities (not least the fact that a bird in the hand is proverbially worth several in the bush). In other cases, advertisers may engage in direct attempts to trade off one criterion of mate choice against another: men who lack resources are more likely to offer cues of commitment and to express a willingness to accept children from a previous relationship (something that men are normally so reluctant to do that most dating agencies advise women advertisers not to mention if they already have children).

Once a couple meet, of course, it is much harder for them to make claims they cannot justify – although the billions spent each year on cosmetics, beginning as long ago as Cleopatra herself, is evidence enough of attempts to exaggerate genuine cues of attractiveness or to stall the inevitable processes of ageing. These genuine cues of attractiveness are studied within the framework of sexually selected traits that play an important role in the sexual and social biology of non-human animals. Sexual selection is the second type of selection that Charles Darwin proposed. Whereas natural selection is directly concerned with phenotypic variation in both males and females and enhances their ability to survive and reproduce, sexual selection acts to increase the relative attractiveness of individuals as potential mates. Sexually selected traits are those that are attractive to members of the opposite sex and, to the extent that they become a basis for mate choice decisions, they are likely to be selected. Unlike naturally selected traits that enhance an individual's survival prospects, sexually selected traits increase the likelihood that an individual might reproduce – sometimes at the expense of their ability to survive. The classic example of a sexually selected trait is the peacock's train: its size and conspicuity makes it easy for predators to detect a peacock and difficult for him to escape predation. However, a peacock's reproductive success is directly linked to his train: females find them attractive and preferentially mate with males which have larger trains with more symmetrical eye-spots. These kinds of male ornamentation are common across the animal kingdom and most of them have been interpreted as the outcome of female choice operating on male traits through sexual selection. Is there any evidence that sexual selection has operated on aspects of human morphology?

One such trait is height which, in males, is certainly attractive to females. Not only are taller men perceived as having higher status, they really do enjoy occupational advantage, as measured by both remuneration and status. Women perceive them to be more desirable dates and taller men (self-reportedly) date more. In addition, it has been suggested that taller men suffer fewer health problems (for example, heart attacks). So, if women do indeed prefer taller men, then it should follow that taller men do better, reproductively speaking, than shorter men. It seems that they do. The few studies that have tested for an association between male stature and reproductive success have consistently demonstrated that, on average, taller men are more likely than shorter men both to marry and to reproduce.

Although women's preference for taller men may reflect a choice for gene quality (stature is, as we pointed out in Chapter 3, a trait with relatively high genetic heritability), it is by no means obvious that it is driven by good genes. It may equally be selection for the man's current or future earning potential (given that, in both traditional and post-industrial societies, there is a correlation between a man's stature and his wealth) or for the man's family wealth (wealthier parents do tend to produce taller offspring, irrespective of genetics). It might even be selection for a good mother-in-law: if she's done it once with her own sons, she might prove to be an effective provider of childcare a second time around. Unfortunately, we do not at present have the data to test these explanations.

If there is a large heritable component to attained height, why don't we see more seven-foot men in the general population? The answer lies in the fact that height is only advantageous up to a point. Beyond a certain stature, the benefits are outweighed by the costs: very tall men experience increased risk of musculo-skeletal diseases and lower back conditions. Daniel Nettle's study of a cohort of British men revealed that the tallest men in his sample (those in the 10th decile of height) were significantly more likely to have work-impairing long-term illnesses. Another factor that might act against unconstrained directional selection for height is the degree of sexual dimorphism in height that men and women find attractive. One recent study found that, whilst both men and women do prefer partnerships in which the male is taller than the female, there is a limit to the degree of height dimorphism that people consider to be attractive. Moreover, there may be constraints imposed by selection on the women themselves that prevent men's height getting out of hand simply because, ultimately, they share the same genes for stature. Daniel Nettle found that, unlike males, women who were at or just below average height enjoyed the highest reproductive success. Women who were taller and shorter than average had significantly fewer offspring than women of average height. This kind of stabilizing selection on women inevitably constrains men's capacity for growth.

We noted above that youth and physical attractiveness are correlates of fertility in women. There is a problem, however, in that it's not always straightforward to pin down exactly what is attractive at any given time. Standards of attractiveness seem to change across time and cultures: what was considered attractive in the 1990s might not have been in the 1970s. In order to demonstrate that a measure of attractiveness does represent an underlying biological function, it's necessary to show a link between variation in that characteristic and measures of fitness. Devendra Singh from the University of Texas at Austin has argued that female body shape, particularly the waist-to-hip ratio (WHR), serves as a reliable (and therefore honest) predictor of fertility. The WHR, measured as the size of the waist divided by the circumference of the hips, reflects patterns of fat distribution that are controlled by sex hormones, particularly oestrogen. Both men and women rate healthy young women with a WHR of 0.7 as the most attractive. Interestingly, since the 1920s, winners of *Miss America* pageants have had WHRs that deviated very little from 0.7 and so too have *Playboy* centrefold models.

It needn't necessarily follow that a high attractiveness rating for a particular WHR means that WHR is a signal of fertility. However, as it turns out, it does: WHR is associated with measures of both fitness and health. Women with a more tubular, male-like shape (WHR approximately equal to 1.0), experience more difficulty in getting pregnant, and have a later age at first live birth than women with lower WHRs. Similarly, since the risk profile for diseases such as diabetes, heart attack and stroke varies as a function of the distribution of fat rather than of the total amount of fat, WHR is also a good predictor of health. Women with low WHRs are at lower risk of developing major health complications. By inference, this suggests that low WHR women should not only be able to produce children but also provide genetic resistance to disease.

While Singh's claims have received some cross-cultural support, other studies failed fully to replicate his findings and claim that the preference for low WHR is an artefact of the pervasiveness of Western culture. In addition, Martin Tovée and others have argued that Singh's analyses didn't adequately control for the effects of body size and that body mass index (BMI), which is a measure of body fat based on both weight and height, is a better predictor of male ratings of female attractiveness. We should make clear that Tovée's study didn't refute Singh's findings but rather claimed that, while both BMI and WHR are predictors of male ratings of female attractiveness, BMI is simply a better predictor. However, a more recent study of Polish women found that BMI was a better predictor of infant birth weight (one good standard for fitness) in small-bodied women but WHR was a better measure in large-bodied women. This might explain why BMI (or body fatness) is a better correlate of attractiveness in traditional hunter-gatherer societies (where women tend to be smaller bodied) but WHR is a better index in industrial societies (where people tend to be larger).

Another region of the body that might be expected to convey honest information is the face. It's well known that faces convey much information about an individual's disposition and play an important part in communication (especially in communicating someone's honesty). Certainly people are able and quite willing to differentiate attractive and unattractive people on the basis of their faces but do those judgements correlate in any way with assessments of the quality of the people being rated? There are a number of facial characteristics that are attractive in both male and female faces (including prominent cheekbones, large eyes and a wide smile) but there are also certain characteristics that are attractive only in men or in women.

Neotenous (or 'infantile') features, which include small chins, small upturned noses and large eyes are rated by men as being attractive in women, whereas females rate prominent chins on males as more attractive. Overall, males find youthful faces more attractive in females, whilst women are reportedly more attracted by mature features in men. The multi-billion pound beauty industry is very much underpinned by a quest to retain youthful looks, particularly by women – and most cosmetic surgery procedures are specifically intended to reduce the signs of ageing, by removing or hiding wrinkles and sagging.

David Perrett and his colleagues at the University of St Andrews have carried out a great deal of work on facial attractiveness. One of the many intriguing results that they have reported is the fact that female preferences for particular male faces vary during the menstrual cycle. Using digitally morphed photographs, they showed that, when women are in the ovulatory (that is, the most fertile) phase of their cycle, they prefer a more masculinized version of a face: they are more attracted to faces that have larger and squarer jaws, high cheekbones and prominent brow-ridges (all of which reflect high levels of the 'male' hormone testosterone). However, during non-fertile stages of their cycle, females express a preference for the feminized versions of the same faces.

One interpretation of these data is that, when they are likely to conceive, females prefer cues that indicate good genes but at other times they prefer cues that suggest the male is less dominant and more likely to invest in a relationship and parenting. This interpretation fits neatly with questionnaire-based studies, which suggest that women prefer heroes for one-night stands but altruists as long-term mates and friends. These findings seem to reflect women's attempts to trade off the different benefits that these two kinds of males have to offer (good genes versus rearing benefits). In an ideal world, they would conceive with males offering good genes and rear the resulting offspring with an altruistic, caring, male.

Since there are only a limited number of males offering good genes, this inevitably creates the conditions for a market in covert matings. Such a market encourages 'caddish' behaviour in males who have the qualities to succeed in these conditions, whilst those who don't (the 'dads') may find themselves making the best of a bad job and tolerating the resulting costs of cuckoldry (providing they sire at least *some* of their mate's offspring). In effect, one sex's attempts to optimize its reproductive performance may influence the mating strategies of the other sex, setting up a series of feedback processes that ultimately generate considerable complexity as they play themselves out.

cryptic clues

Most of the signals of individual quality that we discussed in the previous section are intuitively obvious. But there is also a range of more subtle signals. One of these is symmetry. In organisms that develop bilateral characters (arms, legs, wings, ears, etc.), symmetry is regarded as an indicator of genetic quality, since only individuals with the best gene complexes will be able to grow symmetrical characters in the face of developmental insults such as parasitic infection, nutritional stress and disease. In this sense, symmetry is regarded as an honest signal of both phenotypic and genotypic quality and is widely used in studies of mate choice in species which have developed conspicuous bilateral sexually selected ornaments (for example, forked tails in swallows).

While humans haven't evolved such obvious sexually selected ornaments as a swallow's tail, a number of researchers argue that symmetry is, none the less, an important component of human mate choice. There have been consistent findings that more symmetrical individuals receive higher ratings of attractiveness, are more aggressive and perform better in competitive arenas. Symmetrical males appear more attractive to females and they emerge as better competitors, which might enhance their status and hence their attractiveness. Women also report higher frequencies of orgasm with more symmetrical male partners. High frequency of orgasm probably results in high sperm retention, which would directly impact on the male's fitness.

There is even some evidence to suggest that females are able to smell male symmetry. Steve Gangestad and Randy Thornhill from

the University of New Mexico asked women to rate the smell of
T-shirts that had been worn by symmetrical and asymmetrical men
for two nights. Their results showed that women who were at or near
the most fertile part of their menstrual cycle much preferred the
smell of the T-shirts worn by the symmetrical men, whereas women
who were in the infertile phase of their cycles and women who were
using oral contraceptives showed no preference.

Smell, or the use of olfactory communication, has been relatively
unexplored in humans. For other mammals, chemical communication plays an important part in their social and sexual biology. Mice,
for example, distinguish kin from non-kin by smell, using the *major
histocompatibility complex* (MHC). The MHC genes, which play an
important role in the functioning of the immune system (and hence
create our resistance to parasites), give us our individually distinctive odours and parents and offspring use these odours to recognize
one another (see Chapter 4). Female mice preferentially mate with
males who have MHC genes that are *different* from their own (who
are therefore less likely to be related to them). Claus Wedekind has
shown that normally cycling women also prefer the scent of males
whose MHC is different from their own, whilst women using oral
contraception (which mimics pregnancy) prefer males with a *similar*
MHC. Preferring a male with dissimilar MHC makes sense if a
female is looking for someone to father her children (it provides a
more diverse mix of resistance to parasites), whereas a preference for
similar MHC makes sense if a female is pregnant – yet another
example, perhaps, of *cads* versus *dads*.

A recent study has also shed some light on why humans use perfume. One suggestion was that perfumes mask body odours; if body
odour conveys information about genetic quality, then perfume
users might be able to disguise any shortcomings. In fact, Wedekind
and Manfred Milinski have shown quite the opposite. They examined female perfume preferences and found that their subjects
strongly preferred perfume ingredients that correlated with their
own MHC. From this it might be expected that the perfume ingredients they preferred for themselves would not be the ones they preferred for their partners and there was some indirect evidence that
this was indeed the case.

One more subtle indicator of male quality has been uncovered
by John Manning from the University of Central Lancashire. The
second-to-fourth finger ratio (2D:4D, or the length of the index
finger divided by the length of the ring finger) is highly sexually

dimorphic – in females, the two fingers are approximately equal, but in males the ring finger is generally longer than the index finger. This difference in finger length is believed to be under the control of sex hormones, particularly testosterone, during foetal development and thus provides a measure of masculinization. Manning and his colleagues have shown that low male 2D:4D (that is, high foetal testosterone) is associated with, amongst other things, increased aggression, better sporting and musical ability, higher fertility and possibly higher socio-economic status, while higher female 2D:4D (i.e. lower foetal testosterone) is associated with higher lifetime reproductive success but also higher risk of breast cancer. Manning has also shown that lower 2D:4D ratios are associated with higher female ratings of both male dominance and masculinity, although 2D:4D is not associated with female ratings of male attractiveness. To borrow the title from Manning's book on the subject, digit ratios appear to serve as a pointer to fertility, behaviour and health, each of which is important in male and female mate choice decisions.

This survey of human mate choice tactics is by no means comprehensive. Space has constrained us into focusing on some of its more intensively studied aspects – and we haven't considered the issue of same-sex mate choice at all. It is important to remember that people don't use only a single trait to guide their decisions. Rather, their choices are based on a complex of traits that signal different characteristics. Courtship is a process of negotiation, not unlike a game of poker, in which prospective partners make bids and assess the responses they receive. Although we can define a number of key universal principles that underpin preferences, actual mate choice decisions are taken in a world made complex by the facts that the preferences of the two sexes do not intersect completely and by the presence of rivals. That complexity makes most mate choices best-of-a-bad-job solutions, rather than ideal ones. In the end, heterosexual matings culminate in the production of offspring. It is to this that we now turn.

summary

Evolutionary theory provides a strong explanatory framework for understanding patterns of human mate choice and the characteristics that people prefer in a potential partner arise directly from a sex asymmetry in the costs of reproduction. For females, reproduction is a more costly business than for males and therefore females are expected to be more attentive to cues that emphasize a male's ability and willingness to assist in rearing offspring. By contrast, males are expected to be more attentive to cues of female fertility. Lonely hearts studies of mate choice preferences, as well as studies of actual marriage data, reveal that mate choice decisions do reflect evolutionary considerations. Importantly, the decisions that people make are contingent on how attractive they are to members of the opposite sex. Also, there are a number of evolutionarily salient physical attributes that are believed strongly to influence mate choice decisions and many studies have shown that they are used to influence differential attractiveness decisions.

the dilemmas of parenthood

The starting point of all parental investment decisions is that not all children in a family are treated in the same way; this point is underscored by research findings, which suggest that any two randomly selected, unrelated children will be more similar to each other than will children brought up in the same household. Everyday experience bears this out: siblings are not clones, neither genetically or behaviourally and parents' behaviour towards individual children varies quite dramatically in response to the characteristics of the children, the parents and the social milieu in which they find themselves. One has only to consider parental attitudes as to what constitutes appropriate and acceptable behaviour for sons or daughters. But before we discuss these issues in more detail, we need to define what we mean by the term parental investment.

the costs of reproduction

In 1972, Roberts Trivers defined parental investment as 'any investment by the parent in an individual offspring that increases the offspring's chance of surviving at the cost of the parent's ability to invest in other offspring'. Essentially, in a world of finite resources, any time, energy and resources that are allocated to one offspring cannot be used on a different offspring and therefore, investing heavily in a current offspring will necessarily mean that a parent will be obliged to delay having another. Each such delay, accumulated over several successive offspring, inevitably reduces the total number of offspring the parent can produce in the course of its life.

It should be clear, then, that parental investment decisions are based on an allocation of scarce resources and the basis on which they are made will be determined, at least partly, by characteristics of both the parent and the child. In particular, the reproductive value (defined as the number of future offspring they can expect to produce) of both parent and offspring will loom large in investment considerations. We'll return to this point shortly.

Among mammals, the most obvious form of parental investment is lactation. As long as a female is suckling her offspring, she experiences lactational amenorrhea: her menstrual hormone system is shut down and she cannot conceive. The duration of lactation is one indicator of the amount of investment and studies have demonstrated considerable variation in both the timing and duration of offspring dependence. As we noted in the previous chapter, the human situation is more complex: not only do parents provide for and care for offspring during their immature dependent years, they continue to help beyond maturity, often providing material (and non-material) resources to help their offspring get ahead in the world. For example, parents might pay for further education, they might help finance a home or business, they might help with child-care or, most obviously, they might leave bequests on their death.

These kinds of investment give children a head start and thereby enhance their prospects of eventually attracting a partner and starting a family. You will recall from the previous chapter that females are, on average, more attracted to wealthy or high status males. If parents are able to increase the wealth or status of their sons, then they presumably increase their attractiveness to females. To the extent that parents do invest disproportionately more in sons, then the corollary of that investment is fewer resources available for daughters.

In modern western welfare states, these kinds of investments might not be as obvious as they are in more traditional societies, where services such as education and healthcare are not always free at the point of delivery. In most African subsistence communities, children are an integral part of the household labour force; they herd livestock, collect water and tend crops. As a general rule, education is seen as a means of economic and social emancipation and therefore desirable. However, the cost of educating children is high. Not only do families have to pay school fees and other associated costs but any child who is in school is not available for household labour. Moreover, in some cases, investing in a daughter's education might not be in your own best interests. For example, in patrilocal societies, where daughters live with their husband's family once they marry, it is the husband's family that benefits from the parental investment in daughter's education. Parents are thus confronted by a dilemma: which children, if any, should they educate, and for how long?

a darwinian paradox

Reproduction has such a direct impact on an individual's fitness that the evolution of parental solicitude would seem hardly worthy of comment. Lack of parental solicitude, in contrast, raises Darwinian eyebrows: why should parents limit – or even withdraw altogether – the amount of care they are prepared to offer their offspring? Nowhere is this paradox more stark than in the case of infanticide – the killing of offspring by parents. The natural temptation is to see such behaviour as maladaptive, as reflecting some kind of pathology or disease. On closer examination, however, infanticidal behaviour turns out to be more adaptive than meets the eye. It is a particularly poignant reminder that, when exploring the evolutionary aspects of behaviour, we should not jump to conclusions prematurely nor rely wholly on our gut instincts about what is or is not, what ought or ought not to be considered, 'proper behaviour'.

The solution to the dilemma of infanticide lies in the fact that, by killing offspring, parents might in fact be optimizing the way they distribute their available reproductive investment. Martin Daly and Margo Wilson from McMaster University in Canada have argued that killing offspring can be viewed as 'the desperate decision of a rational strategist allocating scarce resources'. They found that infanticide was widespread (albeit relatively uncommon) in most societies and cultures all around the world. From a cross-cultural review of infanticide, they concluded that there are three main reasons why parents kill their offspring.

The first relates to *paternity certainty* and follows from the simple fact that, until the arrival of DNA-based paternity tests, fathers could never be completely certain about the paternity of their children. Unlike mothers, for whom maternity is 100 per cent certain, there is always a chance that a male did not father the children he is raising. Therefore, in order to avoid investing in children that are not their own (a form of genetic altruism), males are expected to demand assurances from the mother that the child is indeed theirs. Some manifestations of these assurances include the widespread insistence that a bride be a virgin at marriage, as well as the practice, still continued in some parts of the world, where females are claustrated (hidden from the outside world, for example in harems) or chaperoned when venturing from the marital home. Another example of the extreme lengths people go to in order to ensure female chastity is the practice of female cliterodectomy, which is still carried out in parts of sub-Saharan Africa. Such genital mutilation is specifically intended to lessen a woman's sexual interest and is very much linked to ensuring the female chastity on which family honour is partly premised.

Men's concern over paternity seems to underpin the fact that, in many different cultures around the world, both mothers and maternal grandparents are more likely to emphasise father–offspring similarities in the new-born baby than mother–offspring similarities ('Isn't his nose just like his dad's!'). Indeed, mothers are much more likely to comment on offspring resemblance to the father when the father is in the room than when he is not. Whether such similarity is objectively true remains debatable: most babies are pretty much indistinguishable and most of the real physical similarities seem to emerge much later in life. What does appear certain is that there is a strong need for other people to *see* a similarity between father and child and to let the father know that it's there, even if it is not.

The need to assure males of their paternity is not without a basis. Research findings from a number of societies around the world indicate that children who live with at least one non-biological parent suffer disproportionately more than children who live with their natural parents. Daly and Wilson examined Canadian infanticide data and found that children living with a step-parent were 60 times more likely to suffer fatal abuse than were children of the same age who lived with both their natural parents. Furthermore, step-parents were quite discriminating in dishing out abuse. In families where stepchildren and biological children co-resided, adults were much more likely to abuse their stepchildren than their own children. Although there have been some exceptions (for example, Sweden, where special socio-economic circumstances may apply), similar results have been reported from a number of other countries. Among the Ache people of Paraguay, a child whose father has died has a dramatically increased likelihood of dying before 15 years of age. Cause of death is often deliberate infanticide by the man who 'marries' the widow; these males quite openly state that they are not willing to contribute to the costs of rearing other men's children. Along similar lines, Eckart Voland, from the University of Giessen in Germany has shown that, in a historical nineteenth century German population, children whose father had died were highly likely to die *if* the mother subsequently remarried. Crucially, the death of offspring happened before the mother's subsequent marriage. In this case, it seems that young widows were trying to improve their remarriage prospects, because childless widows had significantly better prospects than did widows who had living children.

The second main cause of infanticide that Daly and Wilson identified relates to offspring quality. Setting aside any moral or ethical judgements that we might have, the simple fact is that, in traditional societies, children who are born with severe handicaps or physical deformities are much more likely to be victims of infanticide than are their able-bodied contemporaries. This is particularly true in societies where institutional care of the handicapped is either not available or is prohibitively expensive. In the absence of costly medical intervention, handicapped children are unlikely to reproduce, even if they survive to maturity. Parents are then confronted with the following dilemma: they can invest as many resources as they can afford in attempting to promote the survival of their handicapped offspring (but with the inevitable evolutionary cost that they will almost certainly be a genetic dead end) or they can terminate investment and start again.

Twins offer another example of the same effect. It is not uncommon, in traditional societies, for one or both of a pair of twins to be killed or abandoned soon after birth – remember the archetypal folk tale of Romulus and Remus, the twin founders of Rome, who were abandoned by their parents and raised by wolves. There are two reasons why parents might want to dispose of twins. One is that two babies place an excessive burden on a lactating mother and, by sacrificing one of them, the survival chances and future reproductive prospects of both the mother and the remaining twin are enhanced. Alternatively, but not mutually exclusively, twins tend to be smaller than singletons and birth weight is a strong predictor of a baby's capacity to thrive. Attempting to raise two or even one low quality offspring may simply result in both dying anyway.

The third cause for infanticide identified by Daly and Wilson is that the parents have insufficient resources to maintain investment in offspring. Scarcity of resources features as a factor in both the previous reasons for infanticide. However, even healthy children, with otherwise normal prospects of survival, may be abandoned when the mother lacks the resources to rear them. This point is underscored by the fact that abandoning children to the care of orphanages was common in the southern Catholic countries of Europe in the post-medieval period and reached epidemic proportions in the eighteenth century, especially in the more impoverished cities. For example, in Limoges (France), during the eighteenth century, the number of orphans deposited with orphanages correlated directly with the price of rye, one of many indices of economic hardship (in years when rye was scarce, due to poor harvests, its price rose).

The dilemma that faces every parent in these circumstances is poignantly highlighted by the fact that mothers often attached tokens to the babies they left at the orphanages, so that they could reclaim them later. And yet the chances of their being able to do so were notoriously slim: 98 per cent of babies deposited with orphanages in the eighteenth and early nineteenth centuries died within a year of malnutrition or disease in the horrendously overcrowded conditions. As Sarah Hrdy points out in her book *Mother Nature,* women must have known that they were putting their infants at risk, since the appalling mortality rates in orphanages were a matter of constant public debate. Hoping that they would later be able to reclaim the child somewhat alleviated the guilt of abandonment, for the decision to abandon one's child is never taken lightly, or without intense anguish. But circumstances are sometimes such as to force a parent's hand.

This relationship between resource availability and levels of parental investment is by no means limited only to the historical conditions that prevailed in pre-modern Europe. John Lycett and Robin Dunbar analysed data on contemporary abortion rates in England and Wales and showed that the same considerations influence women's decision whether or not to carry a current pregnancy to term. These analyses revealed that single women were significantly more likely to terminate a current pregnancy than were married women, and this was especially true among younger single women. That this decision reflected the woman's estimate of her future chances of marriage is suggested by the fact that, among single women, the age-specific probability of terminating a current pregnancy was directly related to the age-specific probability of future marriage. Older single women were more prepared to carry a pregnancy to term, since it might represent their last chance to have children before the end of their reproductive lifespan but younger women, with greater opportunities for marriage at a later date, were more likely to opt for an abortion. Significantly, the reverse was true among married women: older married women were more likely to opt for an abortion than younger ones, presumably in order to limit family size and protect their existing parental investment. The findings were confirmed by a similar study carried out using Swedish data.

It is worth emphasising that infanticide is, relatively speaking, rare and operates mainly in exceptional circumstances. What makes infanticide an evolutionary issue is the fact that it occurs at all: it is common enough not to be dismissed as the maladaptive by-product of circumstances. It seems obvious that parents who nurture and protect their offspring will leave behind more descendents than parents who are not particularly attentive to their welfare or fate. But that is not to say that parents are expected to be blind or unresponsive to the environment in which they are raising their children. If the costs of investment are particularly high, relative to the expected returns, then parents who are able to make a blunt assessment and act upon it are likely to leave more surviving offspring than parents who blindly continue investing in offspring who might, for whatever reason, be a genetic dead end. In the harsh economic circumstances of life in traditional societies, these are often stark everyday choices: the decision to opt for infanticide has to be seen in this context and not in the relatively pampered conditions of modern western societies.

when boys and girls are not equal

In a seminal paper published in 1973, the biologists Robert Trivers and Dan Willard argued that parents might, for very sound evolutionary reasons, prefer to invest in one sex of offspring at the expense of the other. Their theory, now known as the Trivers-Willard Effect, was based on the key assumption that maternal condition influences the offspring's ability to reproduce as an adult. If the variance in the reproductive output of the two sexes differs, then it follows that a parent's preference for one sex of offspring over the other will depend on the parent's own condition and the impact this has on the reproductive capabilities of the offspring. Since the sex with the greater variance in output is more risky, parents should only prefer that sex if they can afford to invest enough to ensure that their off-spring come out at the top of the quality distribution. Parents in poor condition, who would only be able to invest a limited amount in each offspring, should opt for the less risky sex.

In mammals which mate polygamously, males typically have a higher variance in the total number of offspring sired in a lifetime than do females. This is a simple consequence of the constraints of the mammalian pattern of reproductive biology, which we discussed in Chapter 4. Females are limited in the number of babies they can produce in a lifetime by the fact each involves a lengthy period of gestation and lactation; males, in contrast, are limited mainly by the number of babies they can sire, and hence by the number of females with whom they mate. This being so, parents who are in better condition should prefer to have sons rather than daughters, whereas parents in poor condition should prefer daughters over sons.

Although Trivers and Willard formulated their original proposal to explain why sex ratios might vary from their normal value of 50:50 male:female, the principle applies equally to how parents distribute their post-natal investment. This would obviously be especially pertinent for mammals, with their heavy burdens of gestation and lactation, and particularly so for humans, who commonly continue to invest in their offspring long after they have been weaned.

Since Trivers and Willard published their theory, there have been a great many tests of their hypothesis in animals. Some have examined sex ratios at birth; others have considered differential post-natal provisioning of offspring. Generally speaking, the results have been equivocal: for example, in a recent review of studies that examined the Trivers-Willard effect in ungulates, only eight of 21 studies, covering 16 species, found evidence that high-ranking females biased investment in sons. Similarly inconclusive findings have been reported for non-human primate studies. One reason for the failure to find clear, unambiguous support for the theory is the fact that it rests on a complex sequence of arguments and rarely are all the data available to check that all the components are upheld.

Attempts to identify the Trivers-Willard Effect in human behaviour have focused on socially stratified societies, where social status or wealth substitute for parental condition. You will recall, from the previous chapter, that wealth generally has a greater impact on male than on female reproductive success: males are able to use their wealth either to enter into polygynous unions or to support larger numbers of offspring but wealth does not necessarily increase the reproductive potential of women in the same way. None the less, opportunities for hypergyny offer daughters from low status families an opportunity to exploit the wealth of higher status families. It follows, then, that wealthy or high status families should bias their investment towards sons rather than daughters, while poor or low status families should prefer daughters whenever doing so enables the appropriate sex to reproduce more successfully.

One of the earliest studies looked at female infanticide in the Indian caste system. Female infanticide was particularly common in the highest castes (the nobility) and has been related to the much reduced marriage prospects for high status daughters. Within the caste system, women from low castes were allowed to marry up the social scale but women from high castes were discouraged from marrying down. As a result, women in the highest castes had a very limited pool of men who qualified as appropriate husbands, whereas high caste sons had an entire social system below them in which to find a bride. In accordance with the Trivers-Willard model, female infanticide was common in high caste families (one high caste family claimed that it hadn't produced a single daughter for over 100 years!) but rather less common in low status families.

A study of the Mukogodo, a hunter-gatherer tribe from Kenya, also provided some evidence in support of a Trivers-Willard Effect: Mukogodo families actively neglect male offspring in favour of daughters. What makes this a possible Trivers-Willard Effect is that the Mukogodo are a client group of the pastoralist Maasai and Samburu tribes, who live in the same area. The pastoralists consider the Mukogodo to be of low status. As a result, Mukogodo girls attract a lower bride price than girls from the two pastoralist groups. However, because they attract a low bride price, the economics of supply and demand make Mukogodo girls an attractive proposition for some of the less well-off pastoralists. Moreover, since pastoralists pay bride price in cattle and other livestock, even a low pastoralist bride price is worth a great deal more to a Mukogodo father than what a Mukogodo man can offer (Mukogodo usually pay bride price in the form of beehives). Mukogodo girls are thus able to marry up the social ladder.

The knock-on effect is a shortage of local brides for Mukogodo grooms, a problem that is exacerbated by the fact that Mukogodo men aren't able to compensate by seeking brides from other tribes, since they cannot afford the higher bride price for such women. As a result, the Mukogodo under-invest in their sons: daughters were far more likely to be taken for medical treatment and for more minor complaints, than were sons. Furthermore, within the same household, severely undernourished sons could often be found alongside their well-fed sisters. The finding that Mukogodo daughters enjoyed higher completed fertility than sons suggests that the strategy of investing more in daughters pays off.

However, many of the studies that have tried to test for a Trivers-Willard Effect should be interpreted with caution. As we pointed out, two of the key assumptions of the theory are that one sex of offspring has a greater variance in reproductive success than the other and that parents differ in their ability to invest (that is, in their condition). If these conditions are not met, then there is no reason why a Trivers-Willard Effect should be expected. Simply, reporting a sex difference in investment is not necessarily the same thing as reporting a Trivers-Willard Effect. More importantly, perhaps, the input (investment) should map on to the output (offspring reproductive success) and this is very rarely reported.

One study that did meet all these conditions looked at parental investment patterns in contemporary Hungary. The gypsy (or Roma) population of Hungary is unusual in that it has been settled in villages since the 1850s. Despite this, they form an underclass, even though they have access to the same economic and social opportunities as ethnic Hungarians. Importantly, however, gypsy women married to Hungarian men have children with higher birth weights, better survival rates and lower frequencies of birth defects compared to gypsy women married to gypsy men, thus confirming one key assumption of the Trivers-Willard Effect. A comparison of two gypsy villages and two ethnic Hungarian villages showed consistent female-biased investment patterns among gypsies and male-biased investment patterns among ethnic Hungarians on a number of different measures (including birth sex ratios, inter-birth intervals, duration of breastfeeding and duration of secondary education – the latter representing a real financial investment because, even under the Communist government, only primary education was free). Crucially, across the four populations, the ratio of investment in daughters relative to that in sons correlated directly with the ratio of the number of grandchildren produced via the two sexes. Each population seemed to be adjusting its investment in the two sexes of offspring in direct proportion to the fitness pay-offs to be gained from them. The importance of this study lies in the fact that it is one of the very few which have demonstrated that biases in investment are associated with the very thing that they are theoretically expected to achieve: an increase in fitness.

born to rebel

In societies which have an economy based on resources that can be monopolized, such as land, it may not be in the parents' long-term fitness interests to divide their resource holdings between several offspring, for doing so very quickly results in farms that are too small to provide sufficient resources for each family. Instead, they may prefer to concentrate their wealth on just one or two of their offspring and leave the rest to fend for themselves as best as they can. In this way, they guarantee the survival of their genetic lineage, since the lucky offspring that inherit the farm have the best possible opportunity of reproducing successfully, whilst anything that the other offspring manage to contribute is icing on the evolutionary cake.

It is important to appreciate that such strategies are a response to economic constraints, rather than a universal form of behaviour. Parents may be expected to opt for primogeniture (the inheritance of the entire parental estate by the oldest offspring) only if excluding the junior siblings is more effective, in net fitness terms, than spreading their resources more evenly. The critical point is when estates become economically unviable if partitioned further. This problem has a long history and many examples can be given. During the early medieval period, for example, partible inheritance (equal shares to each offspring, even if only one sex inherited) was widespread across Europe. However, in the late medieval period, as land became more and more limited in its availability, families began to switch from partible inheritance to primogeniture. By 1400, primogeniture had become the universal norm.

The flexibility of this response is further emphasized by the fact that different societies have solved this problem in different ways but always so as to achieve the same end: maintaining the economic integrity of the lineage's principal reproductive base. In every case, however, the solution (reducing the number of reproductive units within the family to one in each generation) had the consequence of creating a group of disenfranchised young men. Since a disaffected youth group is not a recipe for community harmony, each society had to find a way to defuse what could easily become an explosive situation.

In late feudal Portugal the inheritance crisis produced by the shift to primogeniture among the landed ducal families was handled by encouraging later-born sons to carve out a new empire abroad, thereby instigating the great era of European exploration. This distinction is nicely reflected in where a family's sons died: first-born sons (who inherited the family estate) typically died in Portugal but later-born sons commonly died abroad in Africa, whither they had been sent on expeditions. Interestingly, hypergyny among women and high levels of mortality among younger sons overseas, meant that a large surplus of unmarried daughters began to build up. This was dealt with by placing them in nunneries, where they occupied a privileged position as 'Brides of Christ' – their keep paid for by donations from their families but remaining available, by arrangement, to set aside their vows of chastity should an extra daughter suddenly be needed for a judicious marriage.

In traditional Tibetan society, the solution of choice was to marry off all the sons to a single woman. This had the merit of reducing economic rivalry among the sons (they all inherited the farm) while at the same time providing a guaranteed work force; more importantly, it drastically limited the number of mouths that had to be fed by the family farm (the sons simply could not breed faster than the rate at which their wife was capable). Since the sons married their wife at the same time, sexual rivalry was greatly reduced because the younger sons were often pre-pubertal. Interestingly, the sexually most dangerous son (the second-born) was removed from the arena by putting him, at an early age, in a monastery, where he served a useful social function as a source of communal wisdom and a route of intercession with the supernatural world (about which, more will be said in Chapter 10). Daughters drew the short straw in this system because, with the exception of the fortunate one who married and those who became nuns, they were reduced to a state of virtual slavery on the family farm.

In pre-modern nineteenth century Germany, a rather different strategy was adopted. Here, they developed a system of ultimogeniture (inheritance by the youngest son) which helped to reduce the number of generational transitions each century (important, because the inheriting son had to provide partial compensation to his non-inheriting brothers). Additionally, in order to reduce the number of non-inheriting sons, who would otherwise have been an economic drain on the family landholding, the landed classes (and only the landed classes) practised a strategy that has come to be known as 'an heir and a spare': third- and later-born sons had greatly reduced chances of surviving to adulthood. Indeed, their chances of reaching even their first birthday were about a third less than their two oldest siblings. This was not a consequence of infanticide but rather, like the Mukogodo, one of reduced parental solicitude.

In many respects, the human family unit can be viewed as a system in which parents encourage children to occupy different niches. A particularly dramatic example of this is offered by Frank Sulloway in his book *Born to Rebel*. He examined the personality traits of over 120,000 people and found striking differences between first-, last- and middle-born children. On average, first-born children turn out to be more conformist, conservative and responsible than their later-born siblings, whilst the latter grow up to be more imaginative, flexible and rebellious compared to first-borns. Older children are more likely to take up careers that are associated with conformism, whereas later-born children appear drawn to careers in which non-conformism is more tolerated. First-borns are also more status-oriented than later-born children.

Sulloway argues that these differences emerge as a solution to the problem of scarce parental resources, including parental attention. Initially, first-born children enjoy undivided parental attention. But as soon as a younger sibling arrives, both children have to compete for that attention and even more as successive children are born. Sulloway argues that first-borns develop personality traits that allow them to continue to enjoy the status and attention they had when they were the only child, whilst later-born children have to develop traits that in some way distinguish them from their older siblings, so as to catch the attention of their parents. Since younger children are usually physically smaller than older children, they aren't able to dominate them and thus they are forced to develop behavioural traits that allow them to cope better in the family environment. These developmental differences set children on trajectories that carry though to adulthood.

An alternative strategy for later-born children is to opt out of the competitive environment altogether. Catherine Salmon has suggested that, while early-born children remain more family-oriented, later-born children are more likely to seek alliances with non-kin, such as friends. She found that first-borns were more likely to name a parent as someone they would turn to in times of distress but later-borns were more likely to seek assistance from friends and siblings than parents.

In this chapter, we have outlined events that, taken at face value, seem to contradict the expectations of Darwinian theory. It's hard to imagine how, in a Darwinian world, behaviour like infanticide could evolve. Parents are meant to nurture and protect their offspring, not kill them. However, as we have shown, in exceptional circumstances it might pay parents to do just that, when the longer-term gains offset the shorter-term costs. We should be clear that we're not suggesting that killing offspring is socially acceptable behaviour. It clearly is not. Nor should it be morally justified merely on the grounds that it is (or has been) evolutionarily adaptive. But, there are circumstances where it is probably the best solution to a bad situation and so becomes at least understandable.

What should be apparent from this discussion is the conditional nature of parental investment decisions. There are few invariant rules that apply under all circumstances, other than 'do what you have to do to maximize your fitness'. Consequently, the investment decisions that parents make must take into account a whole range of factors, including the offspring's sex, the socio-economic environment in which families find themselves and the health of both parents and offspring.

summary

Parental investment decisions are based on an allocation of scarce resources. If parents have a finite amount of time, energy or resources, then any that are invested in a particular offspring are necessarily not available to invest in other offspring, either current or future. The decision to invest in a particular offspring will be determined by a number of factors, one of which will be the child's prospects for future reproduction. Children that have little or no prospect of future reproduction are likely to experience underinvestment, or, in extreme cases, infanticide. Perhaps counter-intuitively, this evolutionary perspective raises the possibility that deliberately killing one's own offspring may be rational behaviour under certain circumstances. Importantly, studies of parental investment decisions highlight the conditional nature of human behaviour: parents make decisions about their children with specific reference to the social, economic and ecological circumstances in which they find themselves.

the social whirl

Humans, like all monkeys and apes, are intensely social. It is this sociality that has given primates their evolutionary edge, making them both one of the longest-surviving lineages of mammal (their origins go back to the age of the dinosaurs) and one of the most widespread. Sociality is the consequence of an attempt to cope, in a collaborative fashion, with the challenges of survival and successful reproduction. In most cases, these advantages derive from savings of scale but in the case of monkeys and apes, it is based on a genuine attempt to solve these problems communally. However, by living together, animals inevitably incur costs. The social systems we observe are the outcome of an attempt to balance the costs and benefits of sociality, to trade off one against the other.

In Chapter 4, we saw how human sociality shapes us as children and endows us with the ability both to understand others as individuals as well as to gain an understanding of our culture. In this chapter, we explore in more detail the nature and origins of human societies and consider what is perhaps *the* fundamental problem that they all face: the free-rider.

primate societies

Human societies are complex. This is self-evident from our everyday experiences. Large-scale societies of the kind that characterize the modern world are, however, a relatively recent phenomenon. Towns of even a few thousand individuals appeared only with the neolithic societies of the Near East, around 7000 years ago; cities of more than a million individuals probably date back no more than a few hundred years. For most of human evolution, we lived in small-scale hunter-gatherer societies, characterized by very small, relatively unstable groups, often dispersed across a very large area. It was only with the emergence of agriculture, around 10,000 years ago, that permanent settlements of any size became possible.

Primate social systems, including those of humans, are implicit social contracts where, in effect, members agree to forgo their immediate self-interests, in order to gain greater benefits, in the long run, by solving some ecological problem more effectively. For most primates, this ecological problem will usually be predation risk. By banding together, individuals reduce their exposure to the risk of being caught by a predator, either because they benefit from a 'many eyes' advantage (the time needed to monitor the surroundings for predators can be shared, thus reducing the cost to each individual) or because the presence of many individuals is an effective deterrent to most predators. There is evidence to suggest that, as primate species have colonized more terrestrial and/or more open habitats (where the risk of predation is higher), they have evolved larger groups. In some cases, however, the predators in question may be competitor primate species or even individuals of the same species (for example, males who may commit infanticide).

Group living, however, necessarily incurs costs for its members, simply by virtue of the fact that they are forced into close proximity. These costs typically come in two forms. Direct costs arise as a result of conflict over resources: individuals involved in conflicts waste time and may incur injury. A particularly important form of injury, in this context, is the effect that conflict may have on the endocrine and immune systems: even very modest levels of stress, from casual harassment, can depress the immune system and, in the case of female primates, disrupt the menstrual cycle to such an extent that functional infertility results. Indirect costs arise when one individual takes the resources (food, water, refuge sites) that another could have used, thus obliging the latter to search further afield. One of the most obvious indirect effects in primates is the need for larger groups to travel further each day, often around much larger territories.

During the course of hominid evolution, our ancestors extended their ecological niche and occupied more open habitats. In doing so, they became semi-nomadic on a very large scale: southern Asia, as far east as China, had been colonized by 60,000 years ago, a mere 10,000 years after leaving Africa. The need to cope with increased predation risk almost certainly forced them to live in larger, more co-operative social groups. However, a fully nomadic lifestyle (one that allowed our ancestors to colonize new continents from their African homeland very quickly) probably depended on the ability to share widely dispersed key resources (those that will always be there, however bad the famine or drought). This would have necessitated large-scale exchange networks that covered an area of sufficient size to guarantee that, no matter how bad the drought, there would always be at least one resource depot large enough to accommodate everyone.

Humans share, with their Great Ape cousins, the fact that they live in fission-fusion social systems. These are social groups that are normally dispersed over a wide area, such that only a few members are in physical contact at any one moment. This is particularly clear in modern hunter-gatherer societies: in these, the members of the community are usually dispersed among a number of campsites. Each camp is a temporary home for 25–50 individuals (5–10 nuclear families); individual families may choose to leave and join other camps at any time. However, not every family in the region can join any camp that happens to be convenient. Although casual passers-by may be given shelter, camps normally consist of families (or individuals) from a specific community of 100–200 people, who collectively share rights of access to the resources that their territory has to offer.

The hunter-gatherer community is a virtual group. Although the entire community may gather in one place from time to time – for example, to celebrate coming-of-age rituals or to arrange marriages – this is very much the exception rather than the rule; such events only happen once every year or so. The sense of community that people have comes from knowing who is related to whom (biologically or socially), knowing their individual life histories and knowing that they form part of a specific network of relationships. Even more importantly, perhaps, those relationships are invariably expressed in terms of privileges and mutual obligations.

the social brain

Primates in general, not just humans, have unusually large brains for their body size and this is mainly a consequence of the fact that they have an unusually large neocortex. The neocortex is the thin outer layer of the brain (it is just a few cells deep) within which most of the processes we recognize as conscious thought take place: it evolved in mammals but large neocortices are a primate speciality. Primates have larger brains than other mammals because they have much larger neocortices. The neocortex typically accounts for between 10–40 per cent of total brain volume in other mammals but begins at around 50 per cent in prosimians (the most 'primitive' and mammal-like of the primates) and rises to around 80 per cent in modern humans.

It is now widely recognized that primates' large brains are significantly associated with the distinctive social skills that primates display, an explanation known as the *social brain hypothesis*. The social brain hypothesis suggests that the demands of living in permanent social groups selected for a kind of intelligence that was particularly adept at tracking the relationships that exist between oneself and all the other members of the group and, more importantly perhaps, keeping track of the relationships that the other animals in the group have with each other.

The main evidence to support this hypothesis comes from a series of studies by Robin Dunbar and his co-workers, which showed that relative neocortex volume correlates with various measures of social complexity across the primates. These indices of social complexity include such things as the size of the social group (see Figure 1, overleaf), the size of grooming cliques (or coalitions), the amount of social play, the use of tactical deception (giving false information to mislead rivals) and, in males, the use of more subtle social strategies to undermine the power-based dominance of higher ranking males in the competition for matings. Importantly, for present purposes, they also showed that neocortex volume correlates with the length of the developmental period between weaning and puberty (the period of socialization), suggesting that animals which typically live in larger, more complex, social groups need an extended juvenile period in order to learn and assimilate everything they need to know to manage their social world.

These relationships seem to be specific to the neocortex, and not, by and large, to other sub-cortical regions of the brain. Indeed, they seem to be specific to the more forward (frontal) parts of the neocortex, such as the frontal lobe (the part of the brain associated with what psychologists refer to as 'executive' functions: those processes we associate most closely with rational thinking and

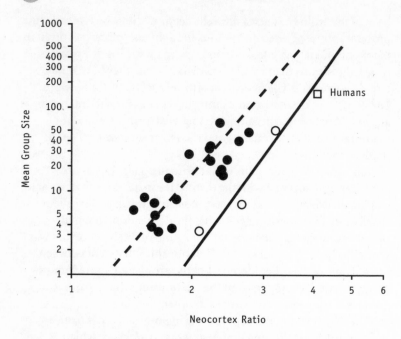

Figure 1. Plotting mean social group size against the species' relative neocortex size (indexed as the volume of the neocortex divided by the volume of the rest of the brain) for different species of primates suggests that there is a simple relationship between the two: in very simple terms, the size of the brain limits the size of social group that can be maintained. Note that ape (open circles) and monkey (black circles) species seem to lie on parallel lines: apes seem to have to work harder for a given group size. The square denotes the size of clans in modern human hunter-gatherer societies. (Redrawn from Barrett et al (2002).)

behavioural control). However, as we noted in Chapter 4, some smaller components, elsewhere in the brain, do also correlate with social group size in primates: these include the size of the parvocellular visual pathway (but not the visual areas as a whole) and one element (but not all) of the amygdala (a part of the ancient limbic system that is concerned with handling emotional cues).

The more rearward (dorsal) regions of the neocortex, such as the visual areas, are largely unrelated to indices of social complexity – despite the fact that the primary visual cortex (usually referred to as area V1) is often the largest single area in the primate brain. This is significant, because, across the primates (including humans), the more frontal regions of the neocortex (the non-V1 regions) have enlarged at a disproportionate rate. In effect, the brain has evolved (and develops in the foetus) from back to front, from the visual areas at the back of the head to the frontal lobe over the eyes. The result is that larger-brained species like Great Apes and especially humans have non-V1 areas and, in particular, frontal lobes that, both absolutely and relatively to the volume of non-cortical brain, are much larger than those of smaller brained species. The significance of this will become clear later.

Humans fit this primate relationship between group size and neocortex size surprisingly well. The typical size of hunter-gatherer communities (around 150 individuals) is exactly the size predicted by the relationship between group size and neocortex volume in primates (and, specifically, the relationship for apes). In traditional horticultural societies, villages typically consist of around 150 people. More importantly, recent studies suggest that this size of social grouping may even be characteristic of post-industrial societies. A study of Christmas card distribution lists, for example, revealed that the number of friends and acquaintances a person has is of about this order too (a mean of 154 individuals for a sample of 42 respondents).

a very social mind

When you plot group size against neocortex volume, one striking feature is the fact that monkeys and apes lie on separate lines (Figure 1). That is, the slopes for the two sets of primates lie parallel, with apes having much smaller groups for a given neocortex size. This is interesting for two reasons. First, the division does not follow simple taxonomic lines: the New and Old World monkeys lie on the same line, despite the fact that the Old World monkeys belong taxonomically with the apes. This suggests that something peculiar happened during the evolutionary history of the apes, quite soon after their ancestors parted company with the ancestors of the Old World monkeys (around 25 MYA). Second, it suggests that apes have to use more computing power to maintain a group of a given size than monkeys do. In other words apes (and, hence, humans) must be doing something more complicated than monkeys to maintain the cohesion of their social groups, and this must have something to do with the complexity of their relationships rather than just with their number.

The one thing that characterizes ape (and human) societies, above all else, is that they have dispersed (or *fission-fusion*) social systems. In effect, apes have to work with a mental world that includes virtual individuals as well as individuals who are physically present, whereas monkeys only have to work with the latter. A plausible explanation is that factoring both present and absent individuals into one's social calculations may be especially taxing cognitively and hence require much more computing power (that is, a larger neocortex).

These cognitive constraints on group size exist as a consequence of our evolutionary heritage. They reflect the demands that natural selection made on our species' sociality during the long hunter-gatherer phase of our existence. This phase was characterized by a form of multi-layered fission-fusion society in which relationships with members of the wider community had to be factored into the relationships with those with whom one happened to be sharing a campsite. This kind of dispersed society may have been critical in allowing our ancestors to adopt a nomadic existence within very large territories. It allowed them to balance the immediate demands of reducing predation risk (by forming temporary hunting camps) whilst at the same time ensuring access to limiting resources on a longer time-scale (through a network of trading relationships).

Our ability to achieve this balance partly stems from the fact that we consider our relationships with others to be real and enduring entities, even in the absence of the people concerned. We think of people even when we don't see them and incorporate them into our lives even when they live on the other side of the world. This, in turn, is linked to our ability to engage in both shared and collective intentionality; our representational abilities (and our language abilities: see Chapter 8) allow us to conceive of mental entities that have no real world manifestation and we understand that people are motivated by beliefs and desires that can be more powerful than and as real as any solid real world object. Given this, it is a small step to view a relationship as a 'solid bond', something that holds people together across space and time, so that constant interaction is not necessary.

circles of intimacy

Neither in modern post-industrial nor in traditional hunter-gatherer societies do we interact with every other member of our community. There is considerable ethnographic evidence (from

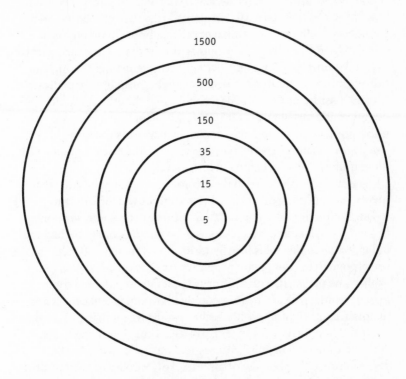

Figure 2. Studies of people's social networks suggest that we each sit in the middle of a series of expanding circles that progressively include more individuals. Each circle demarcates a group of people with whom we have relationships that are of a minimum level of intensity. People who are members of an inner circle mean more to us than those in an outer circle, and we tend to contact them more often. The numbers in each circle (which are inclusive of all individuals in the enclosed circles) seem to be relatively stable, although there is considerable individual variation. The circle corresponding to 150 individuals seems to define the number of people we know individually with whom we have a relationship based on personal trust and obligation; the number 1500 seems to correspond to the size of tribes (those who speak the same language) in traditional hunter-gatherer societies.

both traditional and modern societies) to suggest that the groupings of about 150 individuals that seem to be so characteristic a feature of human social networks are actually structured into a series of hierarchically inclusive subgroupings.

It is as if each of us sits in the centre of a series of expanding circles of acquaintance, with each circle corresponding to a very characteristic number of individuals (see Figure 2, p. 97). These natural groupings seem to cluster at about 5 (the support clique, from whom we would seek emotional or other support in moments of crisis); 12–15 (the sympathy group, with whom we have particularly close relationships); 35 (equivalent in size, interestingly enough, to the typical size of hunter-gatherer overnight camps); 150 (equivalent to hunter-gatherer clans); 500 (recognized in the ethnographic literature as a megaband) and 1500 (equivalent to the tribe, normally defined as the group of people that speak the same language or dialect). (Note that, at each level, the circle includes all those individuals who belong to lower levels.)

The evidence discussed in the previous section suggests that there may be an important distinction between those individuals who belong within the 150 circle and those who lie in the larger groupings beyond. This distinction seems to relate to our knowledge of these individuals *as individuals*. The figure of approximately 150 seems to correspond to the number of people whose relationship to you is explicit and personal, with a history of past interactions and some level of intimacy. These are the people with whom you like to try and maintain contact, in whose life histories you have more than a passing interest. They are the people who, you feel, would be willing to help you with a favour – mainly because there was a sense of obligation between you, either because of some level of intimacy or because of an obligation of kinship or fellowship in an organization or community.

Those who lie outside the circle of 150 we know only as *categories* of individuals: we can label them as belonging to a particular class (say, 'policemen' or 'librarians') and that label provides us with guidelines on how to interact with them. We can recognize many of them as individuals, but we know little about them as people. Our relationship with these individuals lacks the personal warmth that characterizes our relationships with the inner group.

Even within the network of 150 it is possible to see noticeable differences in the intimacy of relationships. In the study of Christmas card distribution lists mentioned earlier, respondents were asked to specify, on a 0–10 scale, how intimate they felt their relationship was with each member of the households to whom they were sending cards. When recipients were ranked in order of intimacy, the total set of recipients tended to cluster rather strikingly at just the kinds of groupings identified above. Significantly, these feelings of intimacy seemed to correlate very tightly with frequency of contact. The sympathy group, of 12–15, for example, seemed to correspond to everyone who was, on average, contacted at least once a month. Interestingly, kinship seemed to have special status in the context of these groupings: kin were significantly over-represented (relative to their availability in the population) within the circle bounded by the sympathy group of 12–15. More detailed study of the size and composition of these various groupings suggests that they may represent real cognitive constraints on the numbers of individuals we can hold in a given degree of intimacy or emotional relationship.

In effect, each circle of intimacy consists of a fixed number of boxes into which we can slot the individuals we meet. Once all boxes in a given circle have been filled, we cannot easily add new individuals. If someone new and exciting comes into town and we want to add them to our social circle at a particular level of intimacy, someone else will have to drop out of that level to make room. Face-to-face contact seems to be crucial in maintaining the quality of the relationship at any given level; failure to maintain contact with someone will have the result of weakening the tie. Old school or college friends, with whom you once spent a great deal of time, gradually become more distant as one grows older. With time, each individual slides inexorably across the circles of intimacy towards the outer edges of mere acquaintance. When you meet up, you can enjoy a few moments of reminiscence but, in all but a very small number of cases (typically those in which the original relationship was one of great intimacy or intensity), a real renewal of the relationship is possible only by starting again from scratch. Your knowledge of them is too out of date, and you no longer have enough in common with them to create a bond of the appropriate intensity. Time, it seems, is another constraint on the number of individuals we can fit into a given social circle.

However, it is clear that the ultimate limit is created by cognitive factors, which influence our ability to maintain coherent and intimate relationships with many individuals. People vary considerably in the size of their social circles at any given level. Social networks, for example, can vary in size from 100–300, even though they have a strong peak at around 150 and similar variance can be seen in the sizes of the more intimate inner circles. In part, these differences reflect a sex difference in sociability: on average, women have larger social networks at any given level than men, even though there is almost complete overlap in the two distributions. In part, it also reflects within-sex differences in personality: there is, for example, a negative relationship between sympathy group size and score on the neuroticism scale of the commonly used MPI personality test: those who score more highly on the neuroticism scale have fewer close friends.

Both these effects owe their origin to differences in social competence and social skills. Evidence for this comes from a study in which individuals were asked to list all the people they contacted at least once a month (one operational definition of the sympathy group) and then tested on advanced theory of mind tasks (the core form of social cognition, which we discussed in Chapter 4). For the test, subjects were presented with a short story detailing a particular social event and asked to identify who was thinking what in the story – with questions that ranged up to nine levels of embedded intentionality. For a sample of 60 subjects, there was a very significant correlation between sympathy group size and achievable level of intentionality (that is, the highest level at which subjects could correctly answer questions about the mind state of the characters in the story). This suggests that social skills and social cognition competency may be directly related – a conclusion that is supported by experimental evidence which suggests that higher order intentionality capacities are lost when individuals are suffering from psychotic conditions such as paranoid schizophrenia or bipolar disorder (manic depression).

trust and reciprocity

The evidence discussed in the previous two sections suggests that our social networks of around 150 people depend on intimate personal knowledge of the individuals included in these circles. That knowledge seems to have important implications for the nature of the relationships involved. It creates a sense of trust and obligation that smoothes the process of interaction – and, in particular, reciprocation and co-operation. The importance of trust and obligation at this level of social organization is emphasized by a number of relevant observations.

The Hutterites, a fundamentalist Christian group, who came from Europe to settle in Dakota and southern Canada during the middle of the nineteenth century, continue, even now, to live a strictly communal life in which the farmland is owned communally and the farm work shared equally. However, they invariably split communities once their size exceeds about 150 individuals because, they say, it is not possible to manage a community that is larger than this by peer pressure alone: you need a police force. Since police forces are anathema to the very concept of their way of life, they prefer to avoid the problem by ensuring that community size is always below the critical limit.

A second example is provided by 'small world' experiments, in which subjects are given a large number of named individuals in different parts of the world and asked to identify someone they know who could be persuaded to take letters to them (passing the letters through the hands of intermediaries if necessary). The results suggest that subjects exhaust their lists of approachable first contacts somewhere between 125–150 individuals. Since this task explicitly asked subjects to request a favour (taking a letter and, if necessary, asking someone else to pass it on in turn), it essentially measures the number of individuals that subjects feel they could ask a favour of without fear of being rejected out of hand or feeling too embarrassed about making such a request. This suggests that a sense of obligation (a willingness to say 'yes' without demanding an immediate quid pro quo in exchange – a definition, perhaps, of true friendship) may be a crucial aspect of the relationships we have with those who are part of our 150 social network.

This kind of evidence suggests that what makes human societies possible is the fact that the members implicitly agree to honour their social obligations. We agree not to steal each other's property, to repay debts – if necessary with interest and not to steal each other's spouses. If we didn't abide by these rules (most of the time, anyway), social life in groups would not be possible. Each family would be forced to space out sufficiently far to avoid risk of conflict or exploitation – indeed, even family life might be impossible, because families themselves are essentially social contracts between couples (and children?).

Trust becomes important in this context, because we do not have the time to check everyone else's reliability and honesty. We simply have to assume that those we meet will abide by the rules. This does not, of course, mean that we are at the mercy of everyone else in society. Much will depend on the dynamic social environment we happen to be part of. History tells us that social life is subject to dramatic mood swings: periods of calm and stability alternate with periods of civil war and chaos. During stable periods, trust and reciprocity grow and we may be willing to act generously towards strangers. But during more troubled times, trust breaks down and we may draw in our social horns to focus our goodwill on the core members of our social networks – those whom we *know* we can trust and on whom we ourselves depend. In climates of suspicion, everyone is looking over their shoulders to check on who is following them. In effect, we are dealing with a predator-risky environment, although in this case the predators are other members of our own species (perhaps even our own community), rather than the more conventional carnivorous kind.

deacon's paradox and the free-rider

The importance of trust emerges in one particular social context, which we call *Deacon's Paradox*. In his book *The Symbolic Species*, Terrance Deacon pointed out that human societies have a peculiarity that exposes them to seriously disruptive forces: that is, humans form pair-bonds, yet have a sexual division of labour. This division means that men and women are often physically separated for long periods of time whilst they are engaged in their different foraging activities. Men and women must agree to honour the integrity of each other's pair-bonds, otherwise the entire fabric of society would collapse.

Yet, as we know from everyday life, other people's relationships aren't quite as sacrosanct as they really ought to be: affairs and illicit liaisons do occur. However, they don't occur with as much frequency as they would if mating were a complete free-for-all, especially given that neither men nor women are around all the time to guard their mates and prevent them forming liaisons with rivals. Moreover, as we suggested in Chapter 5, women may have explicit evolutionary interests in shopping around for 'good genes' even when they are in stable, pair-bonded, relationships, just as much as men may have interests in any sirings they might be able to achieve on the side.

Deacon argued that the capacity to form symbolic contracts, such as formal marriage pacts, has been crucial in the evolution of human sociality. These contracts amount to a formal *public* declaration of commitment that other members of society recognize – and are willing to police. Up to a point, the fact that in traditional societies men and women tend to spend their time apart from their spouses, in single-sex groups where their activities can be monitored, helps to police the system. But neither sex is always under the watchful eye of the other members of society who have a stake in their sexual honesty.

Deacon's Paradox highlights a more general problem, with which any social system founded on an implicit social contract has to contend, namely the destabilizing effect that free-riders inevitably have. Free-riders are those who take the benefits of the social contract but do not pay all the costs. In any social contract, there will always be a temptation to take advantage of the generosity of others. The benefit to the free-rider is often considerable, because they are able to steal a march on everyone else. Consider the classic case, where members of the community exploit a common resource, such as a forest or common grazing land. This resource will last forever if every member of the community uses it sparingly and does not take more than their fair share of the renewable portion of the resource (the proportion that can be replaced by natural growth). However, there is always a temptation to take slightly more – to graze one extra cow on the common or to cut down an extra tree. By doing so, they (and their family) benefit, by having a little extra to see them through the winter. But everyone else pays the cost for their selfishness because there will be less for them to use – the resource will be used unsustainably and eventually will not be available to future generations.

The members of society which the free-rider exploits are forced to behave altruistically: they contribute to the fitness of the free-rider at the expense of their own fitness. These costs may be small in the short term, especially if they are shared between all the other members of the society. But they necessarily add up in the long term. And if the pressures are great enough, the effect of many individuals behaving as free-riders will be such as to impose a very significant burden on the rest of the community. At that point, the implicit agreements that bind the society together will fall apart. Suspicion and a reluctance to engage in reciprocal deals, will increase, making the natural flow of interactions and relationships less fluid. Willingness to co-operate on trust will decrease and gradually the virtual bonds that hold the social system together will dissolve. Free-riding will eventually be held in check by our personal experience of an individual freerider's behaviour: once bitten, we will be reluctant to trust that particular person again. But once we reach that point, the element of trust that helps to hold society together has been lost.

Computer simulations have shown that the free-rider problem becomes increasingly intrusive under two general conditions: when social groups are large and dispersed and when the costs of co-operation are low (that is, when individuals are willing to co-operate without being too inquisitive about whom they are sharing their resources with). Under these two conditions, free-riders will find it relatively easy to locate naïve individuals who are unaware of their behaviour. Both are, as we noted above, features that are particularly characteristic of human social systems.

As with all primate groups, the tensions created through living cheek-by-jowl have to be held in check, lest they overwhelm the benefits and drive the members of the group apart. Primate social groups are held in balance because monkeys and apes can bring to bear sophisticated social cognitive skills that allow them to manage the disruptive forces that act within their societies. These skills are, presumably, underpinned by the computing power of primates' unusually large brains. We explore these processes in more detail in Chapter 11.

summary

Humans are embedded in networks of social relationships that form a series of expanding circles around each individual. Our ability to keep track of the constantly changing world of our social relation-ships depends on the advanced social cognitive capacities that we share with our monkey and ape cousins. The 'social brain hypoth-esis' refers to the fact that primates have unusually large brains com-pared to other animals and that these enhanced cognitive capacities are related to the fact that they have a more complex social life. At the core of this lie the concepts of trust and obligation, which enable individuals to co-operate in groups to solve the problems of survival and successful reproduction in more efficient ways. However, any such system is inevitably plagued by free-riders (those who take the benefits of co-operating, but fail to pay all the costs) and mech-anisms are needed to keep them under control, in order to avoid the delicate balance of relationships in co-operative social systems from being destroyed.

language and culture

Humans are characterized by two features that seem to differentiate them very clearly from all other animals: language and culture. In one sense, drawing a distinction between these two is something of a false division. Culture, in the human sense, depends on language: in the absence of language, human culture would not exist, because language is necessary for the exchange of things cultural. Conversely, language is an integral part of culture: the language that we speak is one very important aspect of our culture. However, it is convenient to differentiate between them in order to consider their evolutionary aspects, because language entails anatomical adaptations for speech that are quite separate from anything to do with culture as such. First, however, we consider how and why human language and culture are unique.

the uniqueness of human being

Over the last century, social scientists have made a great deal of the claim that language and culture are the defining features of humanity. They are, after all, what set us apart from brute beasts. Equally, ethologists have, over the years, been at pains to claim that neither language nor culture are unique to humans. This debate has played such a central role in discussions of what makes humans unique that we need to spend a little time evaluating the claims made by each side. We deal with language first because, historically, it was the first to be explored in any detail.

Language is a system of communication in which arbitrary signs or signals stand for concepts. Conventionally, human language uses an auditory medium, so that language and speech are intimately related. However, as the deaf community demonstrates, human sign languages are as fully functional as are conventional spoken languages. The defining features of language are: the arbitrariness of their signal-meaning relationships (where the sounds or signs used to stand for concepts bear no iconic relationship to the concept itself – the male and female signs used to label public toilets, for example, bear a deliberate iconic relationship to the concept they refer to in a way that the words *male* and *female* do not); the role of grammar in facilitating the coding of complex information; the fact that the sounds are graded rather than discrete (sounds vary into each other rather than being completely different in the way that, for example, a scream and a sigh are categorically different) and the fact that the sounds produced are not, in themselves, emotionally charged (in the way that a scream, for example, most certainly is). This is not to say that emotional overtones cannot be added to human speech sounds but that the nature of the sound itself is not directly caused by the utterer's emotional state: the same sound (or word) can be uttered in any number of different emotional states but still mean the same thing (that is, refer to the same concept or idea).

The traditional view would be that animal communication fails on all these criteria. Their sounds are emotionally charged, have no syntactic (grammatical) structure and do not refer to specific concepts (that is, have meaning or semantics). These claims have, however, been challenged by ethologists. Many species of birds and animals are now known to have different calls for different kinds of predators (technically known as 'reference'). Vervet monkeys, for example, have one kind of call for terrestrial predators like leopards, another for aerial predators like eagles and yet another for snakes. The animals respond to the call with the appropriate evasive action (running for trees, dropping down out of the canopy and standing to peer around into the grass, respectively) they do not need to see the calling animal. Similarly, marmosets are said to have a simple form of grammar that alters the meaning of a particular call.

However, while we can recognize that these claims are all true and that they demonstrate that the precursors of human language are well developed among non-human primates, we are, none the less, obliged to note that no non-human species (not even the honey-bee, with its much vaunted waggle-dance 'language') has a system of communication that is as complex as human language. Recent research has shown that while tamarins (close relatives of the marmoset) can understand simple place-type grammars, they do not understand more complex grammatical rules, which allow sounds that are far apart in the sequence of utterances to be related to each other (something that requires the kind of hierarchical processing of sounds that is required in parsing long grammatically structured sentences). Bees, for example, can tell each other about the locations of nectar sources but that is all: they cannot use their language system to comment on the weather, upbraid lazy drones or discuss where would be a good location for a new hive next summer. So far as we know, such topics of conversation are exclusive to the human species.

Whilst it is relatively easy to define what we mean by language, it has proved much more difficult to arrive at a wholly satisfactory definition of culture. The fact that human behaviour varies so much between different societies makes it seem obvious that it must be cultural in origin. This implies that it has been learned from other members of our immediate social group. This has, however, led to quite different approaches in the three disciplines that take an interest in culture (social anthropology, ethology and psychology).

Traditionally, social anthropologists have understood 'culture' to refer mainly to those aspects of human behaviour which are learned from other members of society. None the less, a famous survey carried out in the 1950s concluded that anthropologists have used the term to refer to more than 140 different kinds of phenomena and that it was all but impossible to identify any one of these as being 'right'. In practice, most of these 140 different usages boil down to just three kinds of phenomena: rules of (usually social) behaviour (for example, rituals, forms of greeting, table manners, etc., including metaphysical and other beliefs about the way the world is); artefacts (things that are made, like tools or structures – what archaeologists usually refer to as 'material culture') and literature, music and art (what we might refer to as 'high culture'). Seen in these terms, it is clear that all three share a property, namely the fact that, in one way or another, they involve ideas in someone's mind. I have a mental image of how you should behave in a particular situation (or why the world is as it is); I have a pattern in my mind when I construct a particular pot or tool; I intend to convey a particular story or meaning when I compose a play or paint a picture. For anthropologists, culture is about meaning and about how that meaning is envisaged.

Ethologists have tended to emphasize the *phenomenon* of culture – the behaviour itself. They note that, when behaviour patterns are learned from others, this can lead naturally to diverging patterns in neighbouring populations. When seeking evidence for culture, they thus tend to place most weight on differences between populations in their habits or styles of behaviour. They have therefore been content to refer to the way that bird or whale songs differ between neighbouring populations (or, within the same population, across generations) as examples of animal culture. When chimpanzee populations in different parts of Africa use different implements for breaking open the hard shells of palm nuts or different types of techniques for obtaining termites from their mounds, ethologists have been content to label these as cultural differences. The populations differ in their styles of behaviour in ways that are not obviously a simple consequence of their local environments. There is no reason why chimpanzees in one population should use grass stems to fish for termites but another population uses twigs.

Psychologists, in contrast, have focused on the mechanisms by which culture is learned. Psychologists point out that behavioural differences can arise through a number of learning processes, some of which we would not want to call cultural. An animal (or, indeed, a human) might have its attention drawn to some aspect of the natural world (say, a food item) by the behaviour of another member of its group. However, if it then figures out how to handle that feature for itself, by trial and error, the result may well be different behavioural traditions in two adjacent populations but we would not want to refer to this as cultural in any meaningful sense. Instead, psychologists insist that the term cultural be reserved for those cases where we can be absolutely sure that an individual is truly imitating what another does, understanding both the object of the act and the means used to achieve it, so that its doesn't need to engage in any trial and error learning in order to work out what to do. When we see evidence for copying of this kind, they argue, then we really can be sure that we have a piece of cultural behaviour.

Anthropologists, ethologists and psychologists thus differ in the criteria they use to identify culture. The first emphasize the social meaning of behaviour, the second the phenomenological variation between populations, whilst the third give central place to the mechanisms of transmission (culture as a process of social imitation). We cannot sensibly say that one definition is more correct than another, because they focus on different questions. We have to decide which task we are engaged in and then use the appropriate definition.

For our purposes, we will understand culture to mean behaviours that are learned from other members of society, and we won't worry too much about the details of the mechanisms involved. Our interest will perhaps be closer to that of the anthropologists: we will focus mainly on the rules of behaviour that underpin human sociality.

how and why language evolved

The conventional assumption has always been that language evolved to allow the exchange of information about the physical environment. Conventionally, this has been interpreted as having, for example, something to do with the organization of hunts, other plans for the future or the giving of instructions (for example, how to make a stone tool). The last decade has, however, seen the emergence of an entirely new suggestion – that language evolved for essentially social purposes. Among the possibilities that have been suggested are the co-ordination of social contracts (something that requires the understanding of symbolic relationships), pair-bonding (the *Scheherezade Effect*, whereby linguistic skills are an honest cue of mate quality and mates use language to keep each other entertained and ensure their continued commitment to the relationship) and social bonding (to facilitate the cohesion of large social groups: the *gossip hypothesis*).

However, whilst all three are plausible possible functions subserved by language, we need to ask whether all three were simultaneously present at the origins of language or if one was the primary function and the others arose afterwards. Whilst any of the three would have significant selective advantages, the gossip hypothesis has the added advantage that it would allow an additional problem to be solved, that is, how to bond large social groups. Thus, social contracts (and, in particular, agreements to respect others' rights to particular mates or marriage partners) may be important for the smooth functioning of society but it is not a problem that is particularly intrusive until you have large social groups, in which there are many rivals. In any case, respecting others' mates or even keeping mates entertained is something that many other species of mammals and birds manage to do without benefit of language. However, once large social groups are in place, the large number of ever-present rivals greatly raises the stakes and social contracts and Scheherazade mechanisms may suddenly come into their own. In contrast, the gossip hypothesis explicitly argues that language was a prerequisite for evolving large groups because it provided the essential mechanism needed to weld them into coherent, stable communities of individuals.

The essence of the gossip hypothesis stems from the observation that monkeys and apes use grooming to bond their social groups. Grooming stimulates the brain to release endorphins (the brain's own painkillers), creating a light 'high'. In some way that we do not really understand at present, the sense of warmth and contentment generated by the flood of endorphins makes monkeys and apes more trusting of, and committed to, the individuals with whom they engage in grooming. Physical contact of this same kind (stroking, rubbing, petting, massaging) has exactly the same effect on us – and we view those with whom we share these activities as special. Physical contact is a mode of communication and one that seems to be particularly capable of high emotional charge. We are able to read a great deal more about the intentions, desires and honesty of the person concerned from a touch than from anything that they might say. A touch is, literally, worth a thousand words.

The critical point in this context is that the time devoted to social grooming correlates with social group size in monkeys and apes. The bigger the group, the more grooming needs to be done by each individual, in order to achieve the same level of social integration. The reasons why this might be so need not detain us here. The important point is that there will inevitably be an upper limit on the amount of time that can be devoted to social grooming and this will ultimately set a limit on the size of group that can be bonded in this way. If a group exceeds this size, it will not be sufficiently cohesive, and will tend to fragment and break up.

This limit on grooming time appears to be about 20 per cent of total day time: the demands of foraging mean that it is not really practicable for animals to devote more than this to social interaction. This sets an upper limit on group size of about 70–80 individuals. But natural human groups average about 150 individuals and if the monkey and ape grooming rates applied to these, we would have to devote around 45 per cent of our time to social grooming. This would be difficult to sustain in the face of the competing demands of foraging. Data from a number of natural human populations suggest that in practice we devote only around 20 per cent of our day to social interaction (the same as the upper limit observed in monkeys) – we use the time we have more efficiently, and that efficiency comes from exploiting a novel medium of communication: language.

Language allows us to use our limited social time more efficiently in at least three, quite separate, ways. One is that it allows us to 'groom' several individuals at a time: at any one time, conversations typically consist of a single speaker with up to three listeners. (If more listeners join the group, the conversation will very quickly split into two separate sub-conversations – unless somebody *very* important is the focus of attention!) In contrast, grooming is very much a one-on-one activity for monkeys and apes: even humans are apt to become affronted if our 'grooming' partner tries to divide their attention between us and someone else. Second, it allows us to exchange information about other members of our network and so keep better tabs on its constantly changing state. For monkeys and apes, what they do not see, they can never know. Third, it allows us to comment on (and so police) the behaviour of others.

when did language evolve?

Irrespective of why language might have evolved, there remains the question of when it evolved. Unfortunately, only bones are preserved in the fossil record, not behaviour, making it difficult to determine just when phenomena like speech or language might have evolved. However, three sources of anatomical evidence provide hints. One is the diameter of the nerve that activates the tongue (or at least the hole through the base of the skull through which it passes). Because speech involves fine motor control of the tongue and lips, it is perhaps no surprise to find that this nerve is (relative to body size) much larger in humans than in other monkeys and apes. Similarly, speech requires fine control of breathing and hence of the chest and rib case; again, humans exhibit a distinctive thickening of the vertebrae in the upper chest compared to monkeys and apes, because they have more nerves devoted to controlling breathing. Examination of fossil hominids suggests that both kinds of neural enlargement occurred around 500,000 years ago, with the first appearance of archaic humans (*Homo heidelbergensis*).

A third source of anatomical insight comes from examining the relationship between group size (as predicted by neocortex size using the primate relationship) and the grooming time required to bond social groups (again, using the standard relationship for primates). If we assume that there is a limit to grooming time above which it is simply not possible to go, we may be able to determine when this happened by plotting the predicted grooming time requirements for different hominid populations against their archaeological date. If we take a threshold at about 30 per cent of time spent grooming as the critical limit (allowing some slippage above the observed 20 per cent in moneys and apes as individuals try to make time savings elsewhere in order to invest in their relationships), we find that this threshold would be breached at about 500,000 years ago, with the appearance of *H. heidelbergensis*. Thus, all three anatomical sources seem to point to about the same time.

More recently, analysis of a genetic difference between humans and apes for a gene that is thought to underpin the capacity for grammar has suggested that the two key mutations (at the FoxP2 gene locus) appeared around 200,000 years ago (roughly when anatomically modern humans, *Homo sapiens*, appeared). This raises the interesting possibility that the capacity for *speech* might have preceded the capacity for *grammar* by some considerable time. How could this be?

One possible reason stems from the observation that speech and language are two quite separate phenomena, as reflected in the fact that some human languages (for example sign languages) do not involve speech. A reasonable interpretation for the apparent temporal separation between the capacity to speak and the capacity for grammatical language is that language was preceded by a phase that used some form of non-linguistic articulated speech for social bonding. The obvious candidate is music – specifically, wordless singing. The possibility that language may have been preceded by a form of communal singing allows us to resolve several puzzles that are otherwise difficult to explain. One is the evolution of music itself and the fact that music clearly has *very* deep emotional valency for us. The second is that it provides a neat bridge between non-human primate vocalizations and human speech that does not require some kind of massive jump: we already find what amounts to wordless song in monkey and ape vocal exchanges. More importantly, perhaps, singing requires exactly the same kinds of vocal control as speech. Stepping up the intensity of song might provide an intermediate step, by putting one of the key prerequisites for language in place. Third, music provides a bridge to language, because it seems to involve the same endorphin charge as grooming (we experience the same sensations of warmth and lightheadedness after singing and especially after communal singing) but at the same time breaks through the grooming threshold: we can sing while we work, and we can sing with more than one individual at a time. Finally, music is clearly phylogenetically older than language: this much is clear from the fact that our musical sense is based in the right hemisphere of the brain (whereas language is based in the left) where it seems to exploit much older, more primitive, neural mechanisms.

darwin, genes and culture

Evolutionary biologists became interested in the phenomenon of culture partly because it promised an opportunity to explore a mechanism of inheritance that was radically different from that governing the inheritance of genetic traits, with which they were more familiar. Conventional biological traits are passed from one generation to the next by a particular mechanism – the molecule known as DNA, via sexual reproduction. However, the possibility that there might exist in the natural world other forms of inheritance, which gave rise to novel evolutionary processes, has always intrigued biologists. One that has excited considerable recent interest is the possibility that viruses might be able to convey genetic material from one individual to another and insert this new material into the genetic code of the recipient individual. Such a mechanism may be responsible for AIDS and it is now being harnessed to create a new medical technology. However, a second mechanism has been familiar to us for a very long time: learning.

On reflection, learning (and in particular social learning) is an obvious Darwinian process: it involves the selective retention and transmission of phenomena (usually rules or ideas). Indeed, the phenomena concerned may even arise in the first place as a novel accident, much like a genetic mutation. While genetic transmission and learning share many key similarities, they are, of course, very different. None the less, setting these important differences aside, they function in very much the same way to transmit information selectively down the generations.

The term *meme* was coined by Richard Dawkins to provide a term which we could use to describe units of culture in the same way we think of units of biological inheritance (genes). You should note that, in this sense, 'gene' does not refer to segments of DNA but rather to what geneticists refer to as *Mendelian genes* (after the founding father of modern genetics, the monk Gregor Mendel, whose contribution to modern evolutionary theory we discussed in Chapter 1). Mendelian genes are phenotypes, or bits of the body (eye colour, a finger, horns) and while they have an obvious relationship to the DNA that codes for them, they are not exactly identical to these chunks of DNA (see Box, opposite). Memes are the cultural equivalent of Mendelian genes, and represent phenomenological units of transmission.

The use of the term meme in this context – and, indeed, the whole idea of a Darwinian process of cultural evolution – has often been criticized by anthropologists (in particular) as being based on an inappropriate analogy with genes and conventional Darwinian evolutionary processes. Genes, so the argument runs, are bits (either of DNA or the body) that can be individually identified but culture cannot be split up into small components (such as individual behavioural rules or the designs on a ceramic pot). Culture is a unitary phenomenon that is passed on from one generation to the next as a monolithic whole. We become members of our community by absorbing its culture (its ideas, beliefs, rules of behaviour) lock, stock and barrel. This might seem a reasonable line of argument but unfortunately it does not really stand up to closer inspection. This is so for three, quite separate, reasons.

First, genes (or even traits, in the Mendelian sense) are not unitary phenomena that can be separated off from the rest of the organism's biology. An arm is a Mendelian trait that is, no doubt, underpinned by some fairly explicit bits of DNA, but it does not make sense to view an arm in isolation from the body to which it is attached – or, indeed, the embryological environment in which it developed. The arm only has biological meaning when it is part of a body, just as a particular rule of behaviour or belief about a particular kind of supernatural being only has social meaning when it is part of the cultural set of a particular society. Geneticists and evolutionary biologists have no problems discussing arms, their evolutionary adaptations and history in isolation. But this does not mean that they are ignoring the rest of the body: the whole is implicit in the argument. In short, this criticism is based on a misunderstanding of what biologists mean when they use a Darwinian approach to analyse biological phenomena.

Second, the anthropologists' view that culture is inherited as a monolithic unitary phenomenon (a view that owes its origins to

MEMES, MENDELIAN GENES AND DNA

The mechanisms of inheritance were first worked out in detail from a series of very careful experiments on peas and other flowering plants, by the Austrian monk and amateur scientist, Gregor Mendel, in the 1850s. He showed that characters (or traits) like pea colour or texture were inherited across generations in a manner consistent with their being passed on by what he termed *factors* (later termed *genes* by early twentieth century biologists). A century later, Mendel's factors came to be equated with DNA, the biological molecule found in every living cell that is responsible for the cell's capacity to reproduce itself.

The strands of DNA found in the nucleus of every cell consist of a series of chemical instructions for making new bodies. However, Mendel's factors are not exactly identical to the segments of DNA that modern biologists also refer to as genes. Mendelian genes are really the characters themselves (eye colour, five-fingeredness, etc.), which are transmitted between parents and offspring with some degree of consistency, whereas the genes on the strands of DNA are more like a recipe to create these characters. Most bits of the body are produced by several separate tiny segments of DNA which can be in different locations on the DNA strand (known as chromosomes), and sometimes even on different strands; equally, some DNA genes can influence several different characters. (That biologists use the term *gene* to refer to several different kinds of biological entity is sometimes confusing to non-biologists; biologists, however, always know from the context just which definition is being used.)

Memes are more like Mendelian genes, in that they are observable elements (rules of behaviour, ideas) that are passed on, more or less intact, from a cultural parent to a cultural offspring. Memes differ from genes in that the mode of transmission is by learning (rather than biological reproduction); in some ways, memes have more in common with viruses and other ways that infections are transmitted. This means that they can involve biologically unrelated individuals (teachers and pupils) and can reproduce themselves much more quickly (learning can be more or less instantaneous). Despite this, however, the same processes of selection are involved.

Emil Durkheim, one of the founding fathers of modern sociology in the early 1900s) reflects a particular historical perspective. Anthropologists (and sociologists) have tended to focus on the here and now of social phenomena. As a result, they ask how the individual acquires the cultural beliefs that it has. The answer is, of course, from its parents, peers and teachers, over a comparatively short period of its early life. Children tend to acquire this information relatively uncritically, believing a thing to be the case or a particular way of doing things to be the right one because they have been told so. But this is a strictly developmental view and lacks an historical (or evolutionary) perspective. We should also ask how a particular culture came to be or why two societies that once had a common origin should end up believing very different things. And it is this second sense that is the focus of Darwinian explanations. How individuals learn their culture is a developmental issue, and a different sense of Tinbergen's *Whys*. We can still legitimately ask how, when and why cultures change over longer time-scales.

Finally, archaeologists studying the historical development (evolution) of Indian basketware and pottery patterns in the historical American West have recently shown that the patterns and elements of basket and pot design tend to segregate through time in parcels. In other words, they are not entirely individual phenomena that can be added or subtracted at will but neither is basket design a monolithic phenomenon that is passed down through the generations. Successive generations adopt or abandon particular suites of characters because they are better in some functional sense *or* because they suddenly become fashionable. These character sets are transmitted through time as cohesive units but they can mix and match at will with any of the other sets of traits that define basket design as a whole. In other words, we can isolate components of a particular cultural phenomenon and analyse its behaviour as a simple Darwinian trait.

One final issue we need to consider is some of the real differences between genes and memes, which have important implications for how we analyse their behaviour from a Darwinian point of view. Genetic traits can – for the most part – only be passed from parent to offspring. Cultural processes, however, exhibit more complex patterns of inheritance. In addition to conventional transmission from a parent to its biological offspring (known as *vertical* transmission), the inheritance of cultural traits can occur between peers (members of the same biological generation: *horizontal* transmission) as well as between biologically unrelated members of the parent and offspring generations (for example, teachers to pupils: *oblique* transmission). This complicates the dynamics of transmission enormously.

Because the mode of transmission is different, it is inevitable that the dynamics of memic transmission will be different. Learning is a naturally faster process of replication than biological reproduction, which has a fixed turnaround time, set by the reproductive cycle of the species concerned. The speed of transmission for a cultural process is limited only by how fast a naïve individual can be found in the population and how long that individual takes to learn the new rule. There is no long delay created by gestation, lactation or the business of socialisation. However, just because learning can make cultural evolution rapid, this does not mean that cultural phenomena will always be extremely labile. Cultural inheritance can sometimes be surprisingly slow. One reason for this is that the heritability of cultural phenomena (that is, the correlation between the behaviour of parent and offspring) is surprisingly high, higher than for many biological phenomena. Children tend to adopt their parents' religion, political views and leisure interests quite reliably but their body weight correlates only about 20 per cent with that of their parents (the rest of the variation in body weight being a result of environmental influences during development). In part, this is a consequence of the intensity of social learning in childhood that we discussed in Chapter 4.

intentionality, language and culture

Hitherto, there has been a tendency to see language as a broadcast activity, much on the analogy of a lighthouse beaming out a message, which is captured by a listener somewhere 'out there'. But this is to miss a key aspect of language, namely that it's a form of *social* communication in which the listener plays as big a part as the speaker. The listener has to make use of considerable mind-reading capacities in order to figure out just what it is that the speaker is trying (*intends*) to say. (Mind-reading is the capacity to understand the contents of another's mind – to see the world from their point of view. We discussed mind-reading or Theory of Mind, ToM and its associated concept of *intentionality* in Chapter 4.) At minimum, the listener has to engage in second order intentionality ('I *suppose* that you are *intending* to mean ...'). If the conversation is about someone else, then third order may be required: 'I *suppose* that you *think* that James *intends*...'.

Theory of Mind is important in language for another reason, to do with some peculiarities in the way we use language. Language can be an extremely precise tool for exchanging information but we often deliberately seem go out of our way to obfuscate. We use metaphors constantly: hardly any word in any human language has only one meaning and most have several that are metaphorical. We speak of water *running*, of people being *rocks* (on which to lean in times of trouble), of *pulling* doors behind us (when we don't mean *pull* in the literal sense of dragging across the road). ToM allows us to sort the metaphorical from the literal meanings, to know when someone is being ironic or sarcastic – and, perhaps most important of all, when they are joking. Joking is a phenomenon that is peculiar to humans and we need ToM to engage in it, as is amply demonstrated by the inability of those who lack it (for example, autistic individuals) to understand both jokes and metaphors.

According to the philosopher John Searle, ToM and language are linked to culture in a profound way. You will recall from Chapter 4 that, as children develop, they move from a shared to a collective intentionality: their growing understanding of their own and other people's minds allows them to understand and absorb the views and practices of their culture. Searle argues that a very large part of this collective intentionality is the creation, understanding of and adherence to *institutional facts*. Institutional facts are facts about the world that exist only because we all agree that they do: marriage, government and money are obvious examples. The particular piece of paper in your wallet or purse is worth five pounds because we all agree that it is: there is nothing inherently valuable about any piece of paper. Searle argues that language and ToM are crucial to the generation of institutional facts and therefore to culture as we know it, because an institutional fact is inherently symbolic and therefore utterly dependent on language. We need ToM to understand that facts of this kind are arrived at by collective agreement and that they exist in everyone's mind in the same way (that is, we all possess the same belief).

These are skills that appear to be unique to humans. Although there is some evidence that chimpanzees (and perhaps other Great Apes) can just about manage the kinds of tasks associated with ToM, their performance is limited to that of children who are on the threshold of acquiring it, and is very inferior to that of even six year old human children, whose ToM capacities are securely developed. And so far as we know, Great Apes are the only animal species capable of giving humans any kind of run for their money in these terms. Monkeys and all other animal species, are limited to first order intentionality. To borrow a phrase coined by Robert Seyfarth and Dorothy Cheney, monkeys are good ethologists (they understand how to read and manipulate others' behaviour) but they are poor psychologists (they don't understand the mind behind the behaviour). Human competencies are not limited to second order intentionality. Normal adult humans are capable of handling problems that involve up to fifth order intentionality, with some exceptional individuals able to perform at higher levels.

There is some evidence to suggest that the levels to which different species can aspire are ultimately dependent on the volumes of core regions of the brain. The frontal lobes of the cerebral cortex may be especially important in this respect, for they are thought to play a critical role in allowing us to separate out reality from fiction as well as to consider alternative scenarios of how things might turn out in the future. If so, then, the relative volumes of these parts of the brain in different species may explain why humans can achieve such startlingly high levels of intentionality, whilst, other species cannot. Analysis of the relative volumes of different parts of the neocortex in primates suggest that it is only with the brain volume of the Great Apes that sufficient spare capacity becomes available to allow individuals to aspire to second order intentionality and this may explain why there is some evidence for ToM capacities in chimpanzees but not in smaller-brained monkeys. ToM and higher orders of intentionality may essentially be an emergent property of the computing power of brains of a certain size and hence of how much neocortical volume can be spared from basic perceptual processing, motor coordination, memory and other conventional cognitive processes.

summary

Language and culture (the capacity to transmit ideas and rules of behaviour from one individual to another through social learning) are unique to humans. Although animals may exhibit both, to some degree, what they have is but a pale reflection of what we find in humans. We suggest that both phenomena are associated with the fact that humans live in large, dispersed social groups that are constantly threatened by free-riders. Aside from the benefits of transmitting knowledge about the world, language provides us with a mechanism for bonding large social groups through a form of 'grooming at a distance'. Not only does language allow us to make declarations of social interest in the people we talk to but it also allows us to exchange information about the state of our social networks, to update ourselves on what is happening and to admonish those who fail to toe the social line. Shared culture, likewise, provides a way of identifying individuals who belong to our social community, those with whom we share obligations and those on whom we can count for moral or economic support when we need it. Both depend on the advanced forms of mind-reading that only humans are capable of and thus probably have a relatively recent evolutionary origin.

the uniqueness of human being

Margaret Thatcher is famous for saying 'there is no such thing as society'. Instead, she preferred to see the world as made up of individuals striving to achieve as much as they could in their own selfish interests. This might sound familiar, since it is the caricature picture presented of Dawkins's selfish gene arguments and forms the basis for the assumption that evolutionary hypotheses are somehow inherently 'right wing'. By now, it should be apparent that just as the selfish gene caricature is wrong, so was Thatcher. Without society, humans would never have achieved their domination of the planet.

However, the other extreme – that it is society at large which determines individual behaviour – is also wrong. Whilst it is true that people show a strong tendency to conform and adhere to the social norms of the culture in which they live (such that we feel as though our behaviour is dictated by culture), it is also the case that, if these social norms prove to be against their interests, people can and do behave in ways that suit them better. This must be true, since otherwise cultures would never change and one does not need to look very far into the past to see that they do.

In the previous two chapters, we have dealt with human social and cultural behaviours, considering how these traits may have evolved, their function and how they enable humans to construct a social reality that has no parallel in any other animal society. In this chapter, we want to consider some of the mechanisms by which cultural changes occur and whether or not these changes are adaptive: does culture help increase an individual's reproductive success?

Or are the effects of culture neutral, the equivalent of icing on the cake – making life a bit more interesting, but unrelated to our capacity to pass our genes on to future generations?

processes of cultural evolution

The intriguing nature of these questions has inspired a number of researchers to build models of the processes by which cultural change occurs and how it is linked to biological evolution. For the most part, these are highly technical mathematical models that do not lend themselves to simple explanations. For that reason, we will not go into them in any detail here (but the bibliography does list relevant sources so that you can follow up these ideas). Essentially, there are three kinds of model, each of which postulates a slightly different relationship between genes and culture.

One model assumes that biology will always keep culture 'on a leash': any trait that proves to have a heavily adverse effect on individual fitness will eventually die out, through natural selection against those who show the trait. In some cases, such as when traits are passed on vertically from parents to offspring, it may be true that the fitness of the cultural trait is directly linked to the biological fitness of the individuals that carry it. However, if looked at from a more interactionist perspective, it becomes apparent that separating learned behaviours (culture) from genetically inherited traits doesn't really make sense: if culturally learned behaviours are inherited like biological traits and have the same fitness effects, then they *are* biological traits. Culture doesn't keep biology on a leash because culture *is* biology.

A second model argues that genetic and cultural traits are not inextricably tied together but are semi-independent. This model aims to discover the conditions under which both genetic and cultural fitness will be maximized. For example, it has been shown that male-biased sex ratios can arise as the result of an interaction between a social preference for sons imposing a strong genetic selection pressure on individuals who produced a disproportionate number of male foetuses. The social preference need not be adaptive as such, but the interaction of cultural practices with heritable biological traits can lead to significant evolutionary change.

The final model treat genes and cultural traits as completely independent. This approach has been termed 'dual inheritance' theory, since it views genes and culture as separate forms of inheritance which may or may not interact with each other. This form of cultural modelling is often seen as the most promising, because it can help explain why we might engage in behaviours that have adverse effects on our fitness.

One thing we need to be aware of is that it may sometimes look as if culture and biology act independently because the cultural under-pinnings of some 'biological' trait are underestimated. Something may appear to be a purely 'biological' trait when in fact it requires a large cultural input for its expression. The cultural roots of a trait can easily be overlooked if they present a pattern which we assume only 'biological' inheritance can produce. As with development, it is the interaction between cultural ideas and biological traits that is important. An obvious example of this is diet. We might suppose that our dietary tastes are dictated by the body's preferences for certain kinds of nutrients and we know that, to some extent at least, this is the case. However, it would be wrong to assume that this is true of all our culinary likes and dislikes. Whether or not we like hot, peppery food, for example, is largely dictated by our experi-ences when we are young. Elizabeth Cashdan has shown that the range of foods that children eat (and, more generally perhaps, even their willingness to try novel foods later in life) is determined by the breadth of their dietary experience during weaning.

When culture acts independently of individuals' biological fitness and especially when this has adverse effects, it is often because there is a horizontal mode of transmission – when ideas are swapped between peers rather than being passed on from parents to off-spring. As we have pointed out, if ideas are passed on vertically from parents to offspring, the success of the learned behaviour may often be dependent on the survival and reproduction of the offspring, who in their turn pass the behaviour on to their offspring. Learned behaviours that are fitness enhancing in the conventional sense will therefore be selected for in the same way as other phenotypic traits and, given enough time, will co-evolve with them. If, however, ideas are passed horizontally, the survival and reproduction of the carrier is less important since the survival of the idea doesn't depend on the carrier having any offspring. Detrimental practices can then flour-ish, because the survival of the individual carrying them is not important to their spread. Dual inheritance theory has thus placed a lot of emphasis on the specific mechanisms by which cultural traits are passed on and learned, because these determine the degree to which genes and culture will interact and co-evolve.

conformity bias and cultural change

One aspect of cultural evolution that is fairly well understood is the psychology of cultural change. The reason people often feel that their behaviour is dictated by their culture is because humans show a strong *conformity bias*: in a given situation, we tend to do what other people do. Even under circumstances where it is clear that other people are just downright wrong, we may 'follow the herd', suddenly losing faith in our own assessment of the situation.

A classic series of experiments, conducted by John Darley and Bibb Latané in the late 1960s, showed just how strong this effect can be. They set up experimental situations where a student was asked to complete a questionnaire about life at university. To do this, the student was then placed in a room with several other 'students', who were really stooges of the experimenters. Several minutes into the experiment, the experimenters arranged for a duct in the wall to begin spewing out smoke – as though there was a fire somewhere in the building. The stooges had been told this would happen and that they should pay no attention but just keep filling in their forms. The naïve student would view the smoke with alarm, look towards the other students, who were calmly writing and, after a moment's hesitation and confusion, would go back to filling in their own form. Occasionally, one of these naïve subjects would get up and inspect the vent but, since no one else had apparently noticed or cared that the room was filling with smoke, they too would often return to their chairs and complete the questionnaire. Over the course of the experiment, only four students left the room and reported the smoke to the experimenter. The rest did nothing at all, even though a fire was apparently raging through the building and putting their lives at risk! In contrast, when a naïve student was placed in a room alone and smoke began to appear, they would report the 'fire' almost immediately. Similar sets of experiments have involved students being asked to make judgements about sets of items. In this case, the stooges would all agree on an answer that was utterly and obviously wrong, for example, that the shorter of two lines was actually much longer than the other. Again, naïve subjects would initially appear confused but nevertheless ended up agreeing with everyone else in the group.

Experiments like these illustrate the conformity bias beautifully, showing that the need to conform can be so strong that people seem willing to die rather than stand out from the crowd. So, what good is it? Something so apparently costly must surely have a counterbalancing benefit, otherwise it would be heavily selected against. Dual inheritance modelling has shown that a conformity bias can be a very powerful way of learning adaptive strategies in 'information-poor' environments, where it is difficult to obtain all the relevant facts by one's own efforts. By adopting what the majority are doing, we average a great number of individual learning experiences and so arrive at a way of behaving that has been tried, tested and proven successful. Allowing the prevailing culture to determine our behaviour can therefore be the smart thing to do. This is even more true when we think of the moral behaviours that help keep everyone in check; conforming to the moral rules of society is essential if individuals are to reap the benefits of living in a community with others. These morals are usually inculcated in us as children and we are often prepared to follow them unthinkingly for the rest of our lives. However, we remain highly sensitive to moral and social codes and can easily pick up those of other cultures, in the main by doing what we see others do. Just as the students did not wish to look foolish by paying attention to a 'fire' and so conformed, we fear making a *faux pas* when in a strange culture and follow the lead of those around us. In doing so, we adopt the moral codes of our hosts, thereby hoping to avoid embarrassment (or worse).

However, if the conformity bias held total sway, new innovative behaviours would never arise. The conformist bias must interact with individual learning and individuals need to be sensitive to environmental change (so that they do not conform to a behaviour that is no longer successful) before significant cultural change can occur. This is why cultures can and do change. Some individual learning always takes place and some individuals are less likely to conform than others. Sensitivity to changes in the environment allows such individuals to respond adaptively bringing in new behaviours that spread through the population by a combination of imitation, other forms of social learning and, once the behaviour becomes widespread, conformity bias.

In some cases of cultural change, the response is produced by a process of co-evolution occurring over several generations but it is also possible for people to show a plastic response during their own lives, adjusting their behaviour to the costs and benefits of new opportunities. For example, in traditional Tibetan society, polyandrous marriage (where one woman is married to several men, usually brothers) is the norm. This was (and to a large extent still is) a response to the harsh ecological conditions under which these particular people live. The poor productivity of their high-altitude habitat means that conservation of the family farm over time was a crucial strategy for the survival of a family lineage, since dividing the farm among inheriting sons in future generations would soon whittle it down to a point where economic survival (and hence reproduction) would become impossible. Polyandry was an adaptive solution to this dilemma, since it limited the number of separate families produced in the next generation to one, whilst, at the same time, ensuring an adequate and co-operative workforce.

However, among the Tibetan populations living in India (such as Ladakh, or Little Tibet), the advent of improved employment opportunities after the 1960s resulted in a substantial reduction in the frequency of polyandrous marriages. As soon as younger brothers were able to become economically independent (for example, in government employment or the tourist trade), they abandoned their polyandrous way of life to make their own independent monogamous marriages (although they often continued to live on the family farm). Even though polyandry was firmly entrenched culturally – and was highly valued – there was little hesitation over changing as soon as the opportunity arose. Of course, sensitivity to changing conditions may not always occur and it is here that cultural practices can lead to fitness losses rather than gains. Later in the chapter, we'll discuss some striking examples of this.

from models to the real world

At present, the field of gene-culture co-evolution is rather heavy with theoretical analyses, like the ones discussed above and light on real data. This is not because the data do not exist but because it is very hard to obtain exactly the right kinds to test the assumptions and predictions of these models. Understanding the interaction between genetic and cultural traits and the evolutionary consequences this can have, remains the greatest challenge for an evolutionary explanation of human behaviour. As we discussed in Chapter 2, the niche constructing activities that humans engage in, through cultural innovation, can lead to unusual and unpredictable evolutionary dynamics and we cannot hope to provide a really comprehensive and satisfying theory of humanity and its origins until we get to grips with this very thorny issue.

In the remainder of this chapter, we will review a number of studies that show how cultural practices and biological tendencies interact. This will give you at least a feel for the kinds of behaviour and practices that are relevant to an analysis of this kind. First, we will present an example of how gene-culture co-evolution can have fitness-enhancing effects. We'll then go on to consider cases where cultural practices are either genetically neutral or explicitly disadvantageous to the fitness of those who practise them.

cultural evolution with functional consequences

One of the clearest and most famous examples of gene-culture co-evolution is lactose tolerance among cattle-keeping human populations, in both Europe and Africa. Although all babies can digest milk, the ability to do so (through the action of the enzyme, lactase) is lost, in most humans, shortly after weaning. Without lactase, the consumption of milk and milk-based products leads to severe diarrhoea, weight loss and, if it continues long enough, death. The only exceptions to this rule are Europeans and a handful of cattle-keeping peoples from the northern parts of Africa. These peoples have retained, as adults, the ability to digest lactose. It seems that, at some time in their evolutionary past, the common ancestor of all these populations acquired a mutation in the gene responsible for producing lactase. Instead of switching itself off at weaning, it carried on producing lactase into adulthood. As a result, these populations could exploit a food type which was previously unavailable. The anthropologist William Durham has argued that this genetic adaptation was reinforced by a behavioural adaptation to exploit milk under conditions of nutritional deficit.

Durham argues that shortages of calcium and vitamin D were the most likely nutritional triggers (the latter may be especially important in northern latitudes, where synthesizing vitamin D in the skin via the action of ultra-violet light is more difficult). Milk is an exceptionally good source of both these nutrients and so the genetic mutation that promotes lactase production would have been selectively reinforced by a cultural disposition to eat dairy products, which in turn encouraged the keeping of domestic stock. If these populations had not been nutritionally stressed and forced to turn to milk for food, the lactase mutation may well have been lost. With no milk in the diet and so nothing to promote its spread, it would have disappeared after a few generations through natural evolutionary processes because individuals with the lactase mutation would have been neither better nor worse off than anyone else.

However, the behavioural response of consuming milk under nutritionally stressed conditions provided a new selective background. Under these conditions, individuals with the lactase mutation had higher fitness than those without, because they could digest milk better and so suffered none of the costs associated with consuming it. The behaviour was therefore essential for the mutation to come into its own and the mutation was essential for the behavioural adaptation to have long-term fitness benefits. Neither would have worked without the other but with both in place, a mutually reinforcing co-evolutionary feedback could be set up and an elaborate and successful culture of herding and pastoralism eventually emerge.

A rather different kind of example is provided by the way languages seem naturally to form dialects. Languages evolve very rapidly and at the level of the local dialect this can be on the scale of generations or even decades. Parent and offspring generations living in the same village can develop noticeably different styles of speech, in terms of both the kinds of words they use and the way these words are pronounced. Daniel Nettle used a computer model of dialect evolution to show that this is probably because dialects provide a very good marker for membership of a social community – precisely because they are learned very young and then remain more or less fixed for life. The model showed that using dialect as a basis for selecting whom to trust in reciprocal exchange relationships was a very effective way of preventing free-riders from destroying communal exchange arrangements.

cultural evolution under neutral selection

As we discussed in Chapter 8, elements of culture – memes – can evolve within their own memetic universe. Memes can be highly successful in terms of their spread through societies but they do not necessarily have any bearing on the fitness of the individuals that carry them, since they spread horizontally rather than vertically and so their fate is not always tied to the reproductive success or genetic fitness of those individuals. Fads and fashions in slang and clothing can probably be seen in this light. There is no fitness cost to wearing combat pants with zips and straps as opposed to zips alone, yet one style may have much higher success in the memic universe than the other.

Fashions or trends can drift at random, since the only thing that determines the success of an idea is how 'sticky' it is – how likely people are to adopt it, relative to other trends. This may well be related to individuals' inherent preference for certain shapes and colours, say, or how well the meme fits with the other beliefs and values that they hold, but the specific form that it takes may not itself be under any particular selection pressure. Since these kinds of traits often involve widespread horizontal transmission between peers, this may explain why certain crazes hit only certain sections of the population and not others. The memic universe of middle-aged men is going to differ from that of teenage girls and so ideas that spread well in the latter are less likely to do so in the former. The advertising industry spends millions of pounds each year analysing the memic multiverses that exist in our culture, trying to infiltrate them and implant ideas that will make specific target groups buy their clients' products. To understand this kind of memic selection, one only has to turn on the television or read a magazine. Our everyday world is a memic laboratory, in which we can all learn something about the processes of cultural evolution.

One excellent example of cultural change that has been driven by an existing human preference is provided by a phenomenon to which we have all probably contributed: the evolution of the teddy bear. When such bears were first invented, during the early 1900s, they were very realistic looking, with pronounced snouts and low foreheads. However, a study by Robert Hinde and Les Barden showed that, as the century progressed, their design became increasingly baby-like and cute, with foreshortened snouts and higher foreheads.

Since the concept of a teddy bear is very obviously not a genetically inherited trait, we can be confident that we are looking at a cultural trait. However, it is a cultural trait that seems to be under the guidance of another, genuinely biological trait: the cues that attract us to babies (high foreheads and small faces). Cute, baby-like features are inherently appealing, producing a nurturing response in most humans. Bears that had a more baby-like appearance – however slight this may have been initially – were thus more popular with customers. Bear manufacturers obviously noticed which bears were selling best and so made more of these and fewer of the less popular models, to maximize their profits. In this way, the selection pressure mounted by the customers resulted in the evolution of a more baby-like bear by the manufacturers.

But who really prefers the baby-like bears – the adults who buy them or the children they're given to? To answer this question, 4–8 year old children were tested for their preference for bears with contrasting features. While there was a tendency for more baby-like bears to be preferred with increasing age, younger children did not exhibit a preference for baby-like features. It seems that, as might be expected, it is adults' preferences that have driven selection for a 'cute' bear, rather than what children themselves like.

when memes go bad ...

One final class of cultural rules concerns instances when they are actively detrimental to the fitness interests of their adherents. One reason for being interested in these examples is that they provide the only certain cases in which we can be absolutely sure that we have an instance of cultural evolution that is independent of conventional natural selection at the genetic level.

Robert Aunger provides us with what is probably the only fully analysed case of this kind. It involves food avoidance practices among forager and horticultural peoples who live together in the Ituri Forest, Congo. Aunger calculated that, on average, for the four tribal groups in his study, food avoidance practices resulted in a shortfall of about 1 per cent in annual nutritional intake (compared to what it would have been without adherence to specific avoidance rules). This is a small effect and comes close to neutral selection, which would allow avoidance rules to persist and evolve by drift.

However, by adhering to their dietary practices, women of the two Sudenaic tribal groups in the sample (the Mamvu and Lese) incurred a measurably significant cost, which equated to a loss of around 5 per cent in lifetime fertility. What differentiates these two particular tribes from the others is that they practise *virilocality*: because women leave their birth homes on marriage (often at a very early age), they lack access to the very people (mothers and older female kin) whom everyone else seems to rely on for information on diet. They thus find themselves in an alien environment, with only their childhood practices to inform them and therefore tend to perpetuate these practices, despite the fact that they may no longer be relevant in their marital locations.

The women fail to adopt a 'when in Rome' conformity bias, apparently because the norms of behaviour they learned as children are just too entrenched to shift. As we noted in Chapter 4, when trait acquisition occurs very early in life (as when individuals grow up in a particular family environment), these traits may be quite robust in their resistance to change later in life; as a result, particular social styles may persist across many generations. Examples include family structures and marriage patterns (monogamy verses polygamy, place of marital residence), kinship terminology and inheritance patterns. The evolutionary biologist Luca Cavalli-Sforza and his colleagues found surprisingly high correlations between the beliefs and habitats of adult offspring and their parents even in the USA.

Such behaviour could also occur because women's food avoidance behaviours are sufficiently similar to those around them that they feel no pressure to conform. It would be interesting to know whether Mamvu and Lese women also fail to conform to other cultural practices within their marital tribes, or whether, when deviation from these practices is more likely to have immediate costs, such as social ostracism, the women adopt them.

At present, we cannot answer these questions, nor do we know whether this effect is stable in the long-term. It could even be that the Lese and Mamvu are currently in the process of learning all the costs and benefits of different dietary habits and that, given enough time, they will change their behaviour as they come to appreciate its adverse consequences. An alternative is that the Lese and Mamvu represent an example of memic evolution at the expense of genetic fitness and if they carry on long enough they will eventually die out as a genetic lineage. Models of cultural evolution developed by Luca Cavalli-Sforza and Marcus Feldman and by Robert Boyd and Peter Richerson both show that, under the appropriate conditions, memes can in fact drive genes to extinction.

An example where the latter clearly happened is given by the Vikings, who maintained a small but viable colony in the southern tip of Greenland for about 400 years, from the last decade of the tenth century AD. At its height, there were around three thousand people living in 280 farmsteads along the south-western coastal fringe; by the twelfth century, the colony even boasted its own bishop and parliament. But, some time after 1408 (when the last ship left Greenland to return to Europe), the colony died out. There is clear archaeological evidence to suggest that the colony finally succumbed to starvation as the Little Ice Age set in, during the middle decades of the fifteenth century: skeletal remains from the cemeteries indicate increasing nutritional stress as the century progressed.

It seems that the Greenland Vikings were unable to let go of their Scandinavian farming practices and adopt the more successful hunting lifestyle of the Inuit. The Inuit were relatively recent arrivals, from the north of Greenland and the Vikings (who lived around the southern fringes) began to come into contact with them during the second half of the fourteenth century. Despite the fact that their farming practices were becoming increasingly less viable, due to the deteriorating climatic conditions, the Vikings seem to have dismissed the pagan Inuit and their culture as beneath the consideration of a (by then) devoutly Christian society. Their outright rejection of Inuit practices on these idiosyncratic grounds meant that the Vikings couldn't reap the benefits that could have come from copying the Inuit – despite the fact that it must have been obvious to the Vikings that the Inuit's practices were extremely successful.

Perhaps their problem was precisely a conformity bias: they came across Inuit people mainly as isolated hunting parties on the sealing grounds around Disko Bay, far to the north of their communities. If the Inuit had come further south, in larger numbers, to live among the Vikings, things might have turned out differently. As it was, the Vikings failed to adopt the Inuit's ecologically more successful lifestyle and the rest, as they say, is history. It seems that they just did not have time to change and adapt to the changing climatic conditions that were overtaking them. Either the pace of climate change was just too fast or they were not sensitive enough to its consequences. Whichever, they simply couldn't adapt their own subsistence practices quickly enough to their new conditions and paid the ultimate price. Meanwhile, the Inuit, with their Arctic-adapted culture, continue to maintain viable populations in Greenland right through to the present day.

We are still very far from understanding exactly how cultural and genetic processes interact to influence the evolutionary process. However, as we develop more sophisticated theories of evolutionary processes that explicitly set out to understand how genetic and non-genetic factors interact, we can begin to formulate hypotheses that specify the kinds of data we need to collect and the kinds of processes on which we need to focus. This is true not only of our own species, but also many other members of the animal kingdom. We are not the only species which learns, and therefore we can expect sources of non-genetic inheritance to be common in animals other than ourselves.

summary

Culture is remarkable for the fact that it imposes an extraordinary degree of consistency on a group of people at any one time. This seems to be a consequence of a conformity bias (a willingness to accept what others in our social community say) that is unique to humans. But cultures do change, and do so on a time-scale that, in evolutionary terms, is relatively short (albeit still in the order of a generation or so). In some cases, cultural change reflects adaptation to new environmental challenges. One example was the adoption of a dairy-product diet (with the genetic changes that make this possible), which seems to have been important in allowing modern humans to occupy latitudes outside the tropics. In other cases, cultural changes may be entirely unrelated to anything in the physical world and may simply reflect social fads that allow us to create a common bond among the members of a particular community.

virtual worlds

In this chapter, we consider two aspects of our social world that are invariably taken for granted: story-telling and religion. Both are unique to humans and both have probably played a crucial role in our evolutionary history. Yet they have been virtually ignored by evolutionary-minded scientists. What makes them particularly interesting, from both the cognitive and evolutionary viewpoints, is that they are concerned with virtual worlds. They deal with the world of the imagination. We noted in Chapter 7 that virtual worlds may be especially taxing cognitively: the demands of having to factor in both individuals who are present and ones who are absent may, we suggested, have been the key factor selecting for unusually large neocortices (and their associated enhanced cognitive capacities) in lineages – like the apes – which are characterized by dispersed social systems. At the same time, we commonly devote large chunks of our time, energy and money to these activities and this should alert us to the inevitable evolutionary question: if something is that costly, it must have a function. But what functions could either of these phenomena possibly serve?

the nature of religion

Religion has always both puzzled and fascinated. It has been a major focus of study in anthropology and sociology almost since the birth of these disciplines, in the late nineteenth century. Early attempts to explore its origins tended to emphasize a relationship between magic and religion and to regard institutional or world religions (those like

Christianity, Islam, Judaism, Hinduism and Buddhism, which
have priesthoods, organizational structures and a formal body of
theology or philosophy) as having evolved from ancestral religions
that were largely based on magic (which involves spells and rituals
designed to achieve particular ends, such as healing, or success in
one of life's many challenges). Religions were, thus, seen as attempts
to understand and/or control the physical world, often through
the intercession of beings existing in a parallel, supernatural,
world.

In a twist on this, Freud (among others) suggested that, in addition to helping us understand the world, religious belief may help
prevent us being overwhelmed by all the adverse experiences we face
in a world that is not always conducive to our well-being and survival
– a world that keeps throwing famines, floods and violence at us with
seemingly boundless enthusiasm; behaviour that can easily be (and
often is) interpreted as malicious and vindictive. Religion provides
hope that the future will turn out better than the present (a claim that
Marx memorably summarized as 'an opium for the people').

Such a view, however, overlooks the fact that religions in traditional societies (and especially those of hunter-gatherers) do not
always have an explicitly magical component. This is not to say that
magic does not exist in hunter-gatherer societies (indeed, it does)
but rather that this magical component is usually quite separate
from religion. In hunter-gatherer societies, religion often has a
shamanic form, in which music, dance and, sometimes,
psychotropic substances, are used to bring about trance states
during which initiates enter into a supernatural world, often in the
company of spirit guides.

Early in the twentieth century, an alternative view (subsequently
widely accepted by social scientists) was advanced by Emil
Durkheim, one of the founding fathers of modern sociology. He
argued that religion was largely designed to reinforce the structure
and integrity of society, by providing a common world-view to which
everyone could sign up, whilst at the same time reinforcing the sense
of belonging – and hence commitment – to the group. Marxists
argued for a slightly different version of this, suggesting instead that
religions (at least in their institutional forms) were attempts by fac-
tions (in this case, social classes) within society to exert control over
society as a whole, by forcing everyone to conform.

There may, however, be more direct benefits to religion.
There is, for example, considerable sociological and psychological

evidence to suggest that actively religious people suffer less illness, have fewer psychiatric problems, recover faster from both illness and surgical interventions, are more contented and generally have a more positive attitude towards life's experiences than do non-religious people. In this respect, it has long been recognized that belief in the efficacy of a treatment (notably the treatments of traditional or explicitly religious healers) is a fundamentally important component of success (the placebo effect).

Whilst all these explanations have merit (and may even all be true), they leave open what must be the most puzzling issue, from an evolutionary point of view – the fact that humans seem to be so susceptible to this kind of exploitation. All else being equal, we would expect any organism to resist attempts to browbeat it into the levels of social conformity that seem to be associated with religions. What is it about religious beliefs and rituals that makes them so attractive that they can be used to persuade and cajole us into signing up to beliefs and commitments that seem quite unreasonable – implausible, even – in the cold light of day? Some authors have been so puzzled by this phenomenon as to suggest that religion is just an *epiphenomenon*: a non-adaptive (perhaps even maladaptive) by-product of something else (such as having a big brain). This view, however, does not make evolutionary sense: anything as costly as religion *must* convey a significant adaptive advantage or the cognitive mechanisms that underpin it would be heavily selected against.

The cognitive underpinnings of religion have, until recently, remained largely unexplored but various experimental studies by the anthropologists Pascal Boyer and Scott Atran suggest that religious claims have some rather particular characteristics. They have to be counter-factual, yet not too implausible. There also has to be an element of the supernatural involved, something which everyday experience tells us is impossible for the normal individual (such as the ability to pass through solid walls or walk on water), otherwise the claims being made seem insufficiently convincing. However, attributions of this kind cannot be made for just any object: an attempt to claim that an inanimate object has these properties is likely to be greeted with disbelief. The supernatural is explicitly associated with living matter. Rocks or rivers can only be credited with believable supernatural powers if they are first imbued with the characteristics of living matter.

This is curious, because it means that believers have to switch off their everyday experience of what is possible (their folk knowledge of physics). That there should be such active suspension of belief implies that the benefits of doing so really must be considerable: were one to ignore the physics of reality in one's everyday life, one would not survive very long.

religion, ritual and the brain

To understand the evolutionary benefits of religion, it is helpful to consider the role of ritual. Although anthropologists have long been fascinated by ritual and its role in society, they have often failed to notice one peculiarity: that many aspects of religious ritual (and especially those associated with religious ceremonial) are often extremely good at triggering the release of endorphins in the brain. Examples include the adoption of painful poses (kneeling, lengthy processions and the meditative poses of yoga), rhythmical movements (dancing, the kinds of rhythmic bobbing exhibited by Jews at the Wailing Wall in Jerusalem, the counting of rosary beads), singing and, sometimes, trials of endurance. In extreme cases, the deliberate infliction of outright pain forms a central part of some rituals: examples include the medieval flagellants who walked from one town to another whipping themselves in emotionally charged ceremonies that drew large and enthusiastic crowds; the self-whipping (and in some cases, slashing with knives) associated with the Islamic Shia cult of the Imam Husain and the self-immolation of the late medieval Khlysty and Skoptzy sects of the Russian Orthodox Church (the former of which survived to the end of the nineteenth century).

All these activities are extremely good at stimulating the production of endorphins. (They may also promote the production of oxytocin and other neurotransmitters as well, although at present we do not really understand what these neurochemicals do, other than the fact that they all seem to be involved in this process and may have much the same kind of effect.) Endorphins may, at least indirectly, have a beneficial effect on the immune system: promoting a sense of psychological contentment may have positive effects on the body's resistance to disease and injury.

The traumatic nature of some ritual acts seems to be very good at inducing the kinds of psychological state that make it easier to 'brainwash' people into believing what you want them to believe. Thus, if one wishes to instil certain rules and values in people, working them up into a religious frenzy first is an excellent way to guarantee that you'll make a lasting impression. Intense emotional arousal can also lead to the formation of strong emotional bonds and the shared nature of many religious rituals (in particular, initiation rituals and rites of passage) is notable in this regard.

Trance states seem to be an almost universal feature of religions all over the world. They may occur spontaneously (as they do in the trances associated with many saints in the Christian tradition) or they may be deliberately brought on by activities that are specifically designed to trigger them – including meditation (in the Buddhist and yogic traditions), rhythmic or repetitive movements such as singing and dancing (trance-dancing in the !Kung San bushmen of southern Africa, as well as in other shamanistic religions and the Pentecostal forms of Christianity) or even pain and the use of hallucinogens (for example, the sweat houses of Native Americans). These forms of shamanistic religion, which are so characteristic of modern hunter-gatherer peoples are now widely thought to be the ancestral form of all human religions.

The neuropsychology of trance states has become a topic of especial interest recently, with a particular focus on the brain areas that are active (or inactive) as adepts enter trance states. In some experiments, experienced meditators were asked to pull a string (as a signal) when they entered a full meditative trance; slow-acting radioactive markers were then remotely injected intravenously and the subject later brain-scanned, after emerging from the trance. It is thus possible to identify which brain areas were particularly involved at the moment they entered the trance.

Entry into trance states is universally associated with a particular set of experiences, including a blinding flash of intense white light, a suffusion of the whole body with a sense of calmness and peace and often a sense of the soul, or self, detaching from the body (variously described as hovering outside the body, flying through space or even entering a completely separate spirit world – interestingly, these also characterize near-death experiences). The brain scan studies suggest that these phenomena are due to oxygen starvation of the left posterior parietal lobe (just above and behind the left ear), which then triggers a feedback loop between the hypothalamus and the attention areas in the frontal cortex. As the cycle builds, it shuts down the spatial awareness neuron bundles, triggering the blinding flash of light and sense of spiritual detachment so characteristic of entry into trance states. However, the sense of calmness that also occurs at this point has all the hallmarks of endorphin release and it seems likely that this is just another example of humans' remarkable ability to find ways of inducing endorphin highs.

Why should endorphin highs be so important (and seemingly universal) an aspect of religious behaviour? The answer seems to hark back to the role of grooming in primate social bonding: you feel positive towards those with whom you share these experiences. It is a very direct form of social bonding, closely analogous to the role that laughter seems to play. However, before we embark on a more detailed consideration of the adaptive function of religion, we need to explore the nature of that other form of virtual world activity, story-telling.

the story-teller's art

On the face of it, story-telling could not seem to be more different from religion. We experience it mainly as a form of entertainment; the books we read, the plays we see. But story-telling plays an important role in religion. At the heart of all the world's religions are collections of stories. Many are origin stories, which explain how it was that the community came to be special; others are exhortatory: they explain the origins of and justification for, the religion's prac- tices. However, outwith the religious element, story-telling has a long and honoured tradition in every human culture. Many of these are origin stories too or recount events that are inspiring in them- selves – the travails and triumphs of real or mythical individuals in the face of adversity.

Like the accounts associated with the great religious writings, they share an additional property: they concern events or people that are deeply meaningful to the listeners. They help to bind the group. They do this partly on the intellectual level, by reminding us why we are a community but also on a more basic level, because our response is often one of pleasure, even laughter – and laughter, as we noted in Chapter 8, is very good at triggering the release of endor- phins. Stories bind on several levels.

It is, perhaps, important to appreciate that stories of the kind we associate with modern literature – psychological biographies, in which we see the world through the mind of one or more of the characters – are of relatively recent origin. Julian Jaynes has argued that they appear for the first time with the Homeric epic the *Odyssey*, which dates to around 800 BC. The earlier Homeric epic, the *Iliad* and all older literature, is characteristically descriptive in form (a narrative that describes events); in contrast, later stories describe the mind states of the protagonists as well as their actions. Although Jaynes equates this – almost certainly wrongly – with the origins of consciousness, in all likelihood it marks an important phase in the long history of our learning how to use language to express our inner thoughts.

Writers' ability to create meaningful and evocative stories probably depends on two key features. One is their ability to observe the hidden reality of everyday life (what makes us tick, what issues really exercise us): stories that do not have some direct relevance to our own struggles with life simply fail to engage. The other is the writers' ability to recognize the audience's cognitive limitations. We will pursue the latter aspect in more detail in the final section but one aspect of it – art as the mirror of the world – will concern us here.

Although attempts to explore the nature of literature and drama from an evolutionary perspective are very much in their infancy, those that have been undertaken agree that the great themes of literature are invariably also the great themes of life – mate choice, parenting, survival, group cohesion and the hero triumphing in the face of adversity.

At a more specific level, analyses of the twelfth century Icelandic Viking sagas have revealed that the events they depict bear a strikingly close resemblance to those we would predict from our understanding of behavioural ecology. In conformity with kin selection theory, individuals are significantly less likely to murder close relatives than more distant ones – unless, of course, the pay-offs for doing so are very significant. Similarly, alliances or loans (of fighting ships, for example) will be made with relatives without expectation of return but strictly on a reciprocated basis with non-relatives and alliances with unrelated individuals are more likely to be reneged on than are those with relatives. One plausible hypothesis (as yet untested) is that the difference between great writers and the more mediocre here-today-and-gone-tomorrow ones is their intuitive feel for the behavioural ecology of everyday human experience.

Perhaps just as important may be the writer's intuitive ability to appreciate the audience's cognitive limitations when following the plot of a story. Analyses of Shakespeare's plays have revealed that his handling of both the size and connectedness of the community on stage mirrors very closely the size of typical human groups. The practical limits seem to be set around fifteen individuals (mirroring the size of natural sympathy groups – see Chapter 7); when the real action of, for example, the Histories requires more parts than this, Shakespeare creates plots within plots to ease the cognitive burden for the audience. Similarly, the number of speaking characters in a scene mirrors very closely the observed size of conversation groups (around four individuals). It seems that the distinction between Shakespeare's more successful plays (*Romeo and Juliet, Othello*) and those which audiences find more difficult (*Titus Andronicus*) rests precisely on how well the number of characters mirrors everyday experience, which is in turn set by the cognitive structures of the human mind.

the role of shared worldviews

The anthropologists' suggestion that religion might be involved in ensuring social cohesion was underpinned by the view (largely derived from Durkheim) that the individual was created by society. In effect, society replicated itself by socializing individuals into its ways. Anthropologists typically interpreted society, in this sense, as meaning culture (the rules and rituals that give society its peculiar form). Whilst this view has been interpreted as a kind of 'good of the species' argument that is unlikely to work under real world conditions, biologists' inclination to dismiss it out of hand merely for this reason may have been premature.

As we suggested in Chapter 9, an alternative view is to argue that society is a collection of individuals who have arrived at some kind of consensus on how people should behave. In this respect, they do impose their views on younger generations through the processes of socialization. But this does not mean that society has an unfettered life of its own: as we showed in Chapter 9, it can and does change, as individual members come to see alternative ways of doing things as being better or more advantageous (to themselves, if not to everyone else). The conflicting views of anthropologists and sociologists on the one hand, and evolutionary biologists and psychologists on the other, derive from the fact that they are dealing with different time frames: the first are essentially concerned with how individuals become socialized, the second with the historical changes that occur within society, on a much longer time-scale.

Recognizing this raises the important possibility that religion (along with other forms of culture) may function to control the disruptive forces which constantly threaten to tear society apart. Given that humans, like all primates, are intensely social and that sociality is the principal basis of their evolutionary success, society is the battleground between each individual's short-term selfish interests and their long-term gains through co-operation. As we saw in Chapter 7, this is the loophole allowing free-riders to gain a foothold, who, if left unchecked, will eventually overwhelm and destroy the essentially co-operative basis on which all societies are founded.

However, foregoing one's immediate interests, in order to gain a greater benefit in the long term, is not that easy and would seem to require high levels of cognitive control; something which may be a distinctly human trait, one which takes a considerable time to develop. As experimental studies of chimpanzees (and young children) have amply demonstrated, Great Apes have considerable difficulty in holding back from an immediate reward, even if by so doing they would gain bigger reward in the future.

In one study, chimpanzees were offered a choice of two dishes, one containing three pieces of fruit, the other seven; the animal had to choose one but the one it chose was given to the experimenter to eat, whilst the rejected dish was given to the chimpanzee. Chimpanzees never managed to master this task, always choosing the dish with the larger reward, even though, trial after trial, they lost out by doing so. The chimpanzees were unable to detach themselves from what was visually present, to distance themselves from the overpowering temptation to satisfy their immediate desires. However, when the food rewards were replaced by cards with the appropriate Arabic numerals, language-trained chimpanzees had no trouble choosing the smaller reward. The abstract symbols of writing apparently gave them just sufficient distance from the reward to allow them to make a more rational judgement.

This can be seen as equivalent to the experimenters providing the chimpanzees with a form of self-control that they cannot naturally manifest. In humans, a small number of regions in the frontal lobe may be critical in allowing us to detach ourselves from the immediacies of actions and their rewards. Thanks to this, we can generate self-control internally, in a way that seems beyond our ape cousins. Among the brain areas involved, the ventro-medial area and the anterior cingulate cortex (ACC) may be especially important. The first seems to be involved in our ability to learn to associate the consequences of particular actions; the second plays a role in allowing us to identify and manage conflicts between perception and presumption (the world as we see it in our heads). The ACC seems to be less well developed in other primates (except, perhaps, the Great Apes) and may be critical in allowing humans to compare and control alternative behavioural strategies.

Given the overwhelming importance of the free-rider problem, we can view religion as a communal attempt to coerce individuals into adhering to the implicit social contract which underpins all societies – the anthropologists are right but for the wrong reason. Religion (and, by extension, story-telling) plays a crucial role in creating a sense of community and bondedness. That effect acts for the benefit of the members, through a group-level effect, because the members of well-bonded groups have higher fitness than those of poorly-bonded groups, making religion a trait that has been selected at the group level. Religion and story-telling are particularly good candidates for group-level selection because they tend to reduce variability between the individuals within groups (since they all come to share the same values and beliefs) and increase variability between groups (which will have different stories and rituals), thereby helping to make the process more powerful.

Since signing up to a religion is a signal of commitment, it is hardly surprising to find that doing so is costly. Evolutionary theory tells us that only costly signals will be honest enough to carry the weight of advertising: if a signal of intent is cheap to perform, it can be cheated too easily. An analysis of nineteenth century American religious communities illustrates this rather nicely; the more sacrifices its members were asked to make (such as giving up tobacco, alcohol, coffee, ownership of property and, even, in some more extreme cases, sex) the longer the commune lasted. Significantly, purely secular communes (which lacked a religious justification for their existence) did not show this effect; despite the fact that they asked similar sacrifices of their members, they had shorter life-spans than communes based on some kind of religious precept. Thus, while the group-level trait of shared religious beliefs enables religious groups to out-compete secular ones, this effect is modulated by the level of sacrifice involved. In religious communities, proof of one's commitment is more convincing if the price of that proof is high.

On face value, religion seems especially well suited to the task of enforcing group solidarity. The imposition of a common worldview creates a sense of communal identity (Us versus Them) that must enhance individuals' willingness to toe the social line. The explicitly moral component of religion clearly reinforces this by providing the threat of punishment from an unseen world. If I try to insist that you conform to the community's view of decent behaviour, your willingness to do so is likely to be a function of whether you see a personal advantage in it. If you happen to think that it is not to your advantage and you don't feel any sense of obligation to those who will be affected by your actions, you can ignore what I ask or challenge my authority. My only real option at this point is physical coercion and I may simply be unable to insist that you adhere to the communal will.

Religion allows those in moral or political authority to circumvent this problem by imposing a threat which it is risky to ignore. The threat of eternal punishment in another life imposes a cost that is difficult to test directly but which – given the time-scale of eternity – it may be foolish to treat cavalierly. It adds a police force to the communal coalition, which no earthly alliance could ever hope to overwhelm. If the risks can be instilled at an early enough age during the process of socialization and involve ritualized brainwashing practices, then the habits of childhood will ensure that the individual always gives more credence to the risks of eternal damnation than is really merited. We will explore the cognitive aspects of this in more detail in the next section.

flights of fancy

In discussing story-telling activities, we alluded to the cognitive loads incurred by both audience and author alike. In this section, we return to this issue and ask how the cognitive loads in both story-telling and religion relate to the natural limits on human social cognition. Let's begin with story-telling.

Perhaps the most common structure for any story is the 'eternal triangle' – the interplay between three core characters (usually the lover, the one desired and a rival). To appreciate the emotional wrench in Shakespeare's classic play *Othello*, the audience has to *realize* [1] that Iago *intends* [2] that Othello *believes* [3] that Desdemona *wants* [4] to love someone else (with the levels of intentionality involved marked out in italics). In any minimally interesting story, the audience has to be able to aspire to fourth order intentionality (the mind states of the three principal characters, plus their own). But in order to compose that story, Shakespeare had to go one step beyond this: he had to *intend* that the audience *realizes*... In other words, the author has be able to factor the audience's perspective on the story into their calculations, as well as those of the characters in the play. This load may be even higher in the action, since, as we noted above, Shakepeare's plays commonly mimic everyday life in the size of the interaction groups he has on stage. Like everyday conversations, scenes typically involve four characters interacting together. To manage this, the audience needs to cope with fifth order intentionality (the four characters plus themselves) and Shakespeare needed to run to sixth order to put the scene together. Since fifth order intentionality represents the upper limits for natural human abilities, this may explain why most of us can appreciate a good story but the ability to write them is surprisingly rare.

We can undertake a similar analysis of the cognitive demands of religion. Whilst we argued in the preceding section that religion carries at least part of its social weight through the psychopharmacological processes associated with endorphins, religion also has a clear cognitive element. Someone needs to work out a point to all these rituals: while the endorphin surge provided by singing or taking part in rituals has its own reward, persuading people to take part in the first place (and, perhaps, especially so in the less immediately rewarding activities) usually requires intellectual justification. That justification is based on the role that some kind of supernatural world plays in our everyday affairs.

In the previous section, we pointed out that one important function which religion may serve is to enforce coercion to the communal will. It is important to appreciate the *communal* dimension to this. It is easy to see that I can have a personal belief in some kind of supernatural world. This requires that I operate with third order intentionality: I *believe* [1] that there is a supernatural being who *wants* [2] me to *act with righteous intent* [3]. Whilst this already requires cognitive capacities that are significantly more sophisticated than anything that any other animal could achieve (Great Apes can probably just about aspire to second order intentionality but there is no evidence to suggest that anything else can get past first order), this is still very much a personal world. There is no reason why anything else in the world should be able to share my momentous discoveries. For this to be possible, you (the audience) need to be factored into the equation, so, at a minimum, fourth order intentionality is required: I *intend* [1] that you *believe* [2] that there is a supernatural being who *wants* [3] us to *act with righteous intent* [4].

So far so good but it is difficult to see how this would force you to conform. At this level, we have shared knowledge but that knowledge is simply descriptive. It is not prescriptive in the sense required for religion to do its job. You may appreciate what I am saying when I try to convince you of the veracity of my claim but why should you feel obliged to agree with me, let alone do as I say? For that to happen, we need to run to one more level of intentionality; we need to move from a purely *social* form of religion to a *communal* one. I have to be able to work at fifth level: I *intend* [1] that you *believe* [2] that there is a supernatural being who *understands* [3] that I *want* [4] him to be *willing* [5] to intervene (when you refuse to conform ...). Hell fire and damnation now provide the necessary stick to the endorphin carrot.

The burden of these analyses is that, in both story-telling and religion-as-coercion, five orders of intentionality are required for those who create the mental worlds involved. Stories and the intellectual structures of religion have something in common: both deal with a fictional world that is not physically present and which can only exist by common agreement – as an institutional fact, in other words. The need to factor a virtual mental world into the physically present one seems to be especially demanding. This seems to provide an explanation for something that is decidedly odd from an evolutionary point of view: why should humans need such cognitively and energetically expensive capacities as the ability to deal with fifth order intentionality. This is puzzling, because everyday experience suggests that we rarely use more than third order intentionality to negotiate our way through the social world.

Fifth order intentionality is costly, because it seems that only brains of a certain size are capable of handling it and brains are, energetically, extremely expensive. Although our brains account for only 2 per cent of total body weight, they consume as much as 20 per cent of the total energy that we consume and their cost is proportional to their size. You have to work extra hard for this luxury, which means that the evolution of super-large human brains must have occurred against a steep selection gradient. Some very dramatic benefit must have been forthcoming to overcome the strong counter-selection created by the costs of large brains. Religion and story-telling seem to be the *only* human activities that require such advanced capacities. If, (as we have argued) these play a critical role in enabling us to bond our super-large groups, then there may have been intense enough selection to evolve these capacities as the ecological need for groups of this size built up during the course of human evolution.

when did religion evolve?

This leaves us with one last evolutionary question: just when did beliefs in a supernatural world and their associated religious activities first evolve? Although the evidence is, at best, sketchy, it seems likely that these capacities evolved at a late stage in human evolutionary history, possibly only with anatomically modern humans. There are three reasons for drawing this conclusion.

First, mapping achievable intentionality levels in monkeys, apes and humans on to neocortex frontal lobe volume (which yields a simple linear relationship) and then in turn mapping this on to the fossil record of hominid brain volumes (using some additional equations) suggests that fourth order intentionality (the minimum for social forms of religion) would not have been achieved until the appearance of archaic humans (*Homo heidelbergensis*) and fifth order, (that required for *communal* religion) would only have appeared with anatomically modern humans (*Homo sapiens*).

Second, archaeologists use deliberate burials (those involving grave goods – everyday artefacts included with the body, presumably for use in the afterlife) as the only certain evidence for belief in a supernatural world. The earliest known burials of this kind date from around 25,000 years ago, although fully articulated skeletons (suggesting inhumation of the body) date from around 50,000 years ago, in the Levant. Although there have been suggestions that Neanderthals may have buried their dead (the evidence for which is now viewed more sceptically), there is no uncontroversial evidence for burials that are not associated with anatomically modern humans as part of the Upper Palaeolithic Revolution. This sets a latest possible date for the origin of religion. As we noted in Chapter 2, the cultural Upper Palaeolithic probably began in Africa, with modern humans, some time around 100,000 years ago. The South African archaeologist David Lewis-Williams has argued that the prehistoric cave paintings which are such a spectacular feature of the Upper Palaeolithic are attempts to represent shamanistic travels in the spirit world.

Finally, since no known modern human cultures lack religion in some form, the capacity for religion must pre-date the emergence of modern humans from Africa, some time around 70,000 years ago. If the cognitive underpinnings of religion had not already existed well before that, we would expect to see much greater variability in this respect among modern humans than we do.

summary

Religion and story-telling are two phenomena that seem to be unique to our species. It is likely that both owe their origins to the need to enforce group cohesion and commitment in the large, dispersed communities that came to dominate human social evolution. In this respect, religion acts as both carrot and stick. Many religious rituals appear explicitly designed to produce the kinds of endorphin surge that seem so prominent a part of the way monkeys and apes bond their social groups. At the same time, the intellectual structures of religion provide both a reason for engaging in these rituals and a threat to those who fail to sign up. The cognitive demands of religion and story-telling seem to be rather similar and significantly higher than anything that even the other Great Apes can aspire to. Mapping social cognitive capacities on to the pattern of hominid brain evolution suggests that the capacities needed for religion and story-telling may not have evolved until modern humans appeared, around 200,000 years ago.

the science of morality

In this final chapter, we want to deal with something that has received almost no attention in evolutionary psychology literature: morality. Why this should be is an interesting question and one which future sociologists of science will undoubtedly find fertile ground for research. One reason, perhaps, is that, when dealing with emotive and controversial issues like infanticide or promiscuity, scientists have not wished to be seen as condoning or justifying socially unacceptable practices and so have insisted that a scientific stance does not equate to a moral one. Yet moral behaviour is clearly something we take very seriously in everyday life. People are prepared to incur considerable costs, including laying down their lives, in the pursuit of a moral dictum. As we pointed out, with respect to religion, in Chapter 10, anything that is so costly must be evolutionarily significant, if only because it has measurable effects on individuals' fitnesses.

We shall try to redress the balance a little here. Our focus will be mainly on the role that moral behaviour plays at the societal level. This fits neatly with the view we have developed throughout this book; that multi-level selection is the key to understanding human evolutionary psychology. Our task has been easier than it would have been even a few years ago because there have recently been a number of studies of just this issue – itself an indicator that evolutionary scientists from different fields are beginning to take these questions more seriously. First, however, we need to explore some philosophical issues.

the naturalistic fallacy

When evolutionary psychologists have sought to get themselves off the morality hook, they have invariably invoked the *naturalistic fallacy*, arguing that just because a behaviour is found to occur 'naturally' or has been selected for, it doesn't follow that the behaviour is somehow 'right' or 'good' (a view neatly encapsulated in the phrase 'ought cannot be derived from is'). However, David Sloan Wilson and his colleagues, Erich Dietrich and Anne Clark, have pointed out that the naturalistic fallacy is really rather a flimsy defence against criticisms concerning the moral import of evolutionary psychology's findings. They argue that not only is it usually misapplied by many (ourselves included) but, when applied correctly, it does not absolve evolutionary psychologists from confronting the ethical and moral issues that the study of human evolutionary adaptations necessarily raises.

So, how should the naturalistic fallacy be applied correctly? As Wilson and his colleagues point out, the Scottish Enlightenment philosopher David Hume, who first enunciated the naturalistic fallacy, did not state that 'ought cannot be derived from is' but that 'ought cannot be derived *exclusively* from is'. The 'exclusively' makes all the difference. Hume's point was that a factual premise cannot, on its own, be used to derive an ethical conclusion; rather, it must be combined with an ethical premise in order to derive that conclusion. Most people misapply the naturalistic fallacy because they exclude the ethical premise; but more importantly, they also fail to realize that the factual premise is essential to draw the ethical conclusion. Thus, while it is indeed inappropriate to say 'behaviour X causes pain and anxiety to others' and from this to conclude 'behaviour X is ethically wrong' without an ethical premise explicitly stating that this is the case ('causing pain and anxiety to others is wrong'), it is equally wrong to conclude that the facts we uncover about behaviour X bear no relation to our moral and ethical systems, since to do so would require that our ethics just appear out of the blue.

This has important implications since, whilst it is true that facts about the world can't be used to justify a particular moral code without some sort of ethical premise to support them, it doesn't follow that the factual statements have no moral force. Ethical principles require facts, if they are to have anything specific to say about how people ought to behave. Appealing to the naturalistic fallacy doesn't side-step moral issues as neatly as many of us have assumed. As Wilson, Dietrich and Clarke so aptly pointed out: 'the naturalistic fallacy cannot be used to ward off ethical debates the way that a crucifix wards off vampires'. That is, there always comes a point where our values must be related to facts about the world and the findings of evolutionary theory are heavily implicated, because evolutionary facts have the potential to change the ethical status of particlar behaviours.

This is already recognized implicitly in our systems of law. As Martin Daly and Margo Wilson have pointed out, English Common Law is based on notions of how a 'reasonable man' could be expected to behave. This is why 'crimes of passion' are often punished less harshly than murders committed in cold blood and people that steal through desperate need are dealt with more sympathetically than those who steal from pure greed.

In some cases, evolutionary findings may render certain behaviours ethically ambiguous, when it might be better, from a societal perspective, if they were deemed totally immoral. Wilson and his colleagues use the example of rape. If it should ever be proved that rape has positive fitness consequences for individuals, then this may have to be factored into our ethical reasoning. If rape harms women physically and psychologically but nevertheless has the effect of increasing the fitness of their offspring (for example, because their sons inherited a tendency to behave in the same way, thereby producing extra grandchildren), this could change the ethical conclusions we might want to draw about this behaviour. If our only ethical premise is that it is wrong to harm another person, then we can conclude that rape is wrong; but, if we also have an ethical premise that it is right to ensure the survival and success of our children (and hence ultimately the success of our lineage), then rape has morally good and bad effects – which makes it harder to reach an unambiguous ethical conclusion. Of course, we can choose to ignore evolutionary facts in our ethical reasoning, if our only concern is to ensure that a particular behaviour is always seen as immoral. The problem is that it doesn't make the facts go away and a reluctance to deal with the point that many natural behaviours are unethical doesn't make for very good science either. In short, real life is philosophically more complicated than naïve folk wisdom would have us believe.

free-riders and the social contract

In Chapter 7, we pointed out that humans' evolutionary speciality, like that of all higher primates, is an unusually intense form of sociality. Such social systems depend on an implicit (or even explicit) social contract, in which individual members accept a short-term cost to immediate personal benefits in order to gain a greater long-term benefit through co-operation. Such social systems are particularly susceptible to free-riders, whose activities risk destablizing the fragile social contracts holding them together. For a species whose fitness depends crucially on the effective functioning of social groups, this is an extremely serious problem. This being so, it is perhaps not surprising to find that we have evolved a number of different mechanisms for identifying and controlling free-riders. Among these are cognitive sensitivities (for example, to social cheating), the use of markers to identify individuals with whom one can afford to take risks (badges of group membership) and the use of reputation to identify free-riders (gossip, in the pejorative everyday sense). In the following section, we explore some of these in more detail. Then, in the final sections, we examine aspects of the pro-social mindset that seem to underpin and support some of these forms of behaviour. Prosociality (behaving in ways, such as generosity, altruism, and forgiving misdemeanours, which seem to benefit the group and its social cohesiveness) seems to be peculiarly characteristic of humans and needs to be explained, precisely because it apparently runs counter to what might be expected in a Darwinian world where (as it is often naïvely misinterpreted) 'dog eats dog'.

evolution's mental firewalls

Predispositions to be sensitive to social cheats have been the focus of considerable interest in evolutionary psychology for nearly two decades. They came to prominence as a result of a series of experiments carried out by Leda Cosmides and John Tooby using the Wason Selection Task. The Wason Task is an abstract logic task, originally developed by the psychologist Peter Wason to study people's intuitive understanding of scientific reasoning. In this task, subjects are given a rule ('Cards with a vowel always have an even

number on the reverse side'). They are then shown four cards like
the following:

A	H	3	4

and asked to say which card or cards they would turn over to test the
validity of the rule.

A logical rule of the kind $P \Rightarrow Q$ (P *implies* Q: the vowel card has
an even number on its reverse) can only be tested by showing that
both $P \Rightarrow \sim Q$ (P *implies not-Q*: the vowel card has an odd number
on its reverse) and $\sim P \Rightarrow Q$ (the card with a consonant has an even
number) are *not* true. Hence, in the above example, the logically correct answer is to check the A and 3 cards: the A card must not have
an odd number on its reverse and the 3 must not have a vowel. It is a
well-established experimental result that only 25 per cent of subjects
answer correctly (the same proportion that would do so by chance).
Most subjects choose the A card alone or the A and 4 cards.

Cosmides and Tooby showed that when tested on a task that was
logically identical but formulated as a social contract ('Only people
older than 18 years are allowed to drink beer: which of the following
four people would you need to check to see if the rule is being
broken – the person who is over 18, the person who is under 18, the
person drinking beer or the person drinking coke?'), 75 per cent of
subjects pick the right answer. It is obvious whom one should check,
because the social rule allows anyone to drink coke and over-18s to
drink what they like. Cosmides and Tooby argued that these results
reflected a specialized cognitive module that was specifically
sensitive to social cheating.

The intrepretation of these findings have been the subject of
intense criticism and debate during the last decade. Many argue that
Cosmides and Tooby's experimental task is confounded by a
framing problem: the particular way in which the task is presented
can lead the subject towards or away from the correct answer. It is
obvious that the causal relationship can only be one way in the social
task (and this makes the answer obvious), whereas there is nothing
to suggest that this is true for the abstract task. Cosmides and Tooby
countered this by showing that the same results were obtained if the
abstract task were presented as a real world non-social task (for
example, a filing task for a clerk): in this case, just as many people

made mistakes as with the original abstract task. However, even here, it is difficult to avoid framing problems and the debate has yet to be finally resolved.

Whilst this dispute continues to exercise minds, everyday experience none the less tells us that people do seem to be particularly sensitive to (or, at least, are unusually concerned about) social cheats. We usually disapprove of such behaviour and can immediately recognize when rules of this kind are infringed. There is experimental evidence which suggests that subjects are more likely to correctly remember events described in a story if they concern some kind of reprehensible behaviour (cheating on the system), as well as evidence showing that we are more likely to remember a person's face if we have been told that they have cheated on some social contract than if we are given neutral information about them. It's as though we are primed to be sensitive to free-riders.

In addition to remembering the facial features of cheats and free-riders, we may also rely on more static cues of group membership. Identifying someone as being a member of the same community gives us with reason to believe that, without needing to know more about them as individuals, there is every likelihood they will co-operate with us out of a sense of obligation – or, at the very least, we can assume that they will operate by the same rules as we do. Group identity is often explicitly displayed by clothing and hairstyles, religion and other beliefs or dialect and styles of behaviour. Wearing certain clothes, speaking in a certain way or having particular kinds of knowledge (how to play cricket, for example) mark us out as belonging to a particular community.

This sense of community may be especially strong when we share the same dialect. The interesting thing about dialects is that they are learned very young: by the time we reach our teens, they are more or less fixed. After that, only a few gifted mimics can learn new ones. Your dialect thus marks you out very strongly as a member of your birth community. Because it is learned young, it is a signal that is difficult to cheat and so is a very reliable cue of community membership. Indeed, until the 1970s, it was possible to place a native English speaker to within 30km of their place of birth simply on the basis of hearing them speak a few words.

Dialects seem to be especially effective at preventing free-riders taking over a social world. This was demonstrated in a computer simulation run by Daniel Nettle. In the artificial world of this simulation, co-operation was critical for successful reproduction, but co-operators were easily exploited by free-riders, who soon took over the population. However, when individuals were only willing to co-operate with those who shared the same dialect (represented in the simulation by a six-digit barcode), free-riders found it much more difficult to prosper. This was most effective when dialects changed rapidly (by more than 30 per cent per generation). This, of course, is something that is especially characteristic of dialects: even within a given location, there are often clear generational differences.

Whilst not an infallible cue to honesty, social markers of this kind at least provide a first estimate as to whether we should be cautious or extend a tentative offer to co-operate. Reliable cues of membership of the same birth community may have an additional advantage, in that they may also identify biological kinship. Sharing kinship in this way allows us to take more risks when behaving altruistically, since investing in individuals who share our genetic ancestry means that the benefits will still accrue to our genetic lineage even if the individual reneges on the implicit arrangement to repay the debt. Because the risks are higher if we decide to accept repayment in the next generation (as it were) rather than in the present one, it is perhaps not surprising to find that cues of natal community membership (such as dialect) are difficult to acquire (and have to be acquired early in life) but cues that identify membership of a community where personal reciprocation would be expected are less difficult to acquire (things that can be learned, like shared knowledge or beliefs that can be acquired as adults).

The last form of mechanism for managing the free-rider problem is the use of reputation. Humans take their reputations very seriously and are often willing to defend them vigorously. This is because spreading the word about another individual's behaviour is a powerful mechanism for controlling free-riders. If we want people to co-operate with us, we must ensure that our reputations for honesty and repayment of debts are snow-white. In another computer simulation of the behaviour of free-riders in a world of co-operators, the Swedish biologists Magnus Enquist and Olof Leimar showed that being able to remember and exchange information about the behaviour of those with whom one co-operated made it much more difficult for free-riders to exploit other members of the community.

In experiments with real people, Manfred Millinski and his colleagues found that, when rounds of 'public good' games (those where individuals must co-operate for the common good, rather than their own personal advantage) are interspersed with 'image-scoring' rounds (where individuals give other individuals a score for how co-operative they perceive them to be), co-operation on the first type of game remains much higher than when there are no reputation-based rounds, suggesting that having to maintain a reputation helps to keep people honest. As we saw in Chapter 8, gossip (in its negative sense) is one of the functions that language makes possible, even though it may not be the only function it serves or even the one that was responsible for its evolution.

evolving an ethical sense

David Sloan Wilson believes that evolutionary psychologists' reluctance to confront moral issues stems principally from the emphasis placed on 'selfish gene' individualistic views of evolution. When adopting a stance where everything is viewed as, at best, enlightened self-interest, unethical behaviours frequently emerge as an evolutionary product. However, ethical behaviours do not seem to 'pop out' in quite the same way. Since it feels rather uncomfortable to have good evolutionary explanations for immoral behaviours when there are no convincing analyses to show that our moral behaviours are also evolved adaptations, there is perhaps a natural tendency to reach for the naturalistic fallacy as a defence. Doing so obviates any need to discuss the morality of behaviour. As might be expected, given his interests, Wilson points out that the problem can be resolved if we adopt a multi-level selection perspective, since ethical or moral behaviours then emerge as adaptations at the group level in the same way that many unethical behaviours do at the individual level.

Understanding that selection can act on individual traits that increase the fitness of groups naturally brings moral behaviour into the equation, as we have argued in the preceding chapters. Language, culture, religion and story-telling are all implicated in the moral issue of controlling free-riders and it appears that the various mechanisms used to control and punish free-riders are just as surely the products of natural selection as the immoral acts they punish. This has raised interest in the evolutionary significance of punishment. Robert Boyd and Peter Richerson used a mathematical model to show that co-operation could be stable under standard Darwinian conditions even in large social groups, providing defaulters were punished. So long as the cost incurred by punishing is less than the gain from co-operation, punishment could be an evolutionarily stable strategy. Indeed, even a moralistic strategy (punishing those who fail to punish defaulters) can be evolutionarily stable.

There is now a large body of experimental evidence to show that people are willing to punish free-riders at a cost to themselves (so-called 'altruistic punishment'), even if there is no possibility that they will engage with the free-rider in any subsequent interaction. Ernst Fehr, an evolutionary economist from the University of Zurich, argues that people are motivated to punish by strong emotions of resentment and annoyance at cheats. It seems as though a desire to see cheats punished, even if they have not cheated one personally, may arise from a sense of 'fairness' which requires that miscreants receive their 'just desserts'. Traditional theories of reciprocal altruism and kin selection cannot adequately account for this response, whereas group-level selection for traits designed to stop free-riders destroying group trust and harmony is a much more promising avenue to explore. Fehr suggests such traits have been selected in humans by a process of multi-level selection at the group level.

strong reciprocity and the prosocial 'instinct'

In a series of experiments, Fehr and his collaborators found that a significant proportion of people willingly repay gifts and punish individuals who violate fairness and co-operative norms, even under conditions where all the individuals remain anonymous, only a single round is played (so-called 'one-shot' games) and everyone is genetically unrelated. Anonymity and the one-shot nature of the games means that reciprocal altruism cannot be operating (individuals cannot recognize their opponents, and never meet them again anyway), whilst the lack of relatedness means these results cannot be explained by kin selection. Fehr has termed this behaviour *strong reciprocity* and defines its essential feature as a willingness to sacrifice resources both in rewarding fair behaviour and punishing unfair behaviour, even if this is costly and provides neither present benefit nor future economic rewards for the reciprocator.

However, by no means all individuals play fair: there is always a significant proportion who pursue a strictly selfish strategy. In contractual games, where one player (the employee) agrees to perform an economic service in return for a reward (a wage) provided by the other player (the employer), employees invariably deliver their service at well below the contracted level; whilst they are happy to agree on a fair contract, the great majority of employees (around 75 per cent!) seem to have second thoughts when it comes to fulfilling it. When offered the opportunity to punish defaulters or reward honesty, around 70 per cent of employers did so. And when employees were then exposed to employers who punished, the number of employees fulfilling their contracts trebled.

In these kinds of games, much seems to depend on whether individuals believe they will be punished for non-co-operation: when they do, they will increase their contributions before any punishment can occur. In games with no punishment, co-operation is not only lower from the start but also tends to diminish over successive rounds. Significantly, the rate of decline is faster when a player is teamed up with strangers in each successive round of the game than when they are teamed up with the same individuals in every round. This suggests that an expectation that you will meet the same people in the future (being in a stable group) does impose some constraints on people's willingness to cheat. This is reminiscent of Mameli's mind-shaping expectancy effects, whereby certain behaviours can be produced purely as a consequence of a person's expectations about another's behaviour.

Willingness to punish is not unconditional. When, in other experiments, the cost of punishing defectors was altered, willingness to punish declined when the costs became too high. Nor is punishment always required to make defaulters toe the social line. In a series of experiments carried out by the political scientist Elinor Orstrom and her colleagues, a group of individuals played anonymously in a co-operative market (that is, one where their returns were determined by how many individuals contributed to a common pool but where individuals choosing to act alone could benefit at the expense of everyone else). Typically, the pay-offs earned by the players in such a game averaged around 25 per cent of what they could have earned had they all co-operated. However, when allowed to meet in a refreshment break part-way through the experiment, the opportunity to harangue the (still anonymous!) defectors was enough to increase mean pay-offs to around 75 per cent of the maximum in subsequent rounds.

To get a better understanding of why strong reciprocators were also willing to punish those who didn't contribute, Fehr and his colleagues decided to ask people why they had punished the free-riders. The answers were interesting, in that practically everyone explained their behaviour in terms of feelings of resentment and anger towards the unco-operative members of their group: the emotional response to this moral infringement was simply too strong to ignore, even though they would have been financially better off if they had held their tempers. Even more intriguingly, when individuals were asked to place themselves in the position of a free-rider and predict the amount of anger other individuals would feel towards them, individuals who had been low contributers when playing the games expected a higher intensity of negative emotion than did high contributors.

Significantly, in the light of our earlier discussions, another factor influencing an individual's willingness to punish in these kinds of games is their perception of their opponent's intentions. In a test that sought to distinguish between unfair *intentions* (where the opponent could choose an outcome) and unfair *outcomes* (where the outcome was determined by a throw of dice), players rewarded and punished opponents significantly more often in the first condition than they did in the second. They seem to take intention into account.

As we noted above, the willingness to co-operate changes very rapidly in experiments where the opportunity or the cost of punishment is varied. The immediate drop in co-operation and punishment seems to be due to the change in the beliefs that the two players hold, rather than being a consequence of trial-and-error learning. This suggests the importance of the ability to control emotional responses according to the cost-benefit ratio of the acts concerned and is further evidence of the subtle mind-shaping effects that drive players' responses. Notice the subtle inferential reasoning that such a situation would require, involving a minimum of three levels of intentionality: 'I *believe* that he *knows* that I *know* that increased costs mean less punishment'.

All this experimental evidence suggests that moralistic altruistic punishment, mediated by strongly negative emotions towards free-riders, is an adaptive response by individuals, and aimed at sustaining co-operation, which, in itself, is a response to the fact that many individuals are willing to co-operate with others in a strongly reciprocal fashion without regard for any future rewards.

Herb Gintis, another evolutionary economist, has mathematically modelled this process, assuming that, first, throughout our evolution human groups would have faced potential extinction threats (floods, famines and other environmental catastrophes) on a regular basis and second, that groups with high numbers of strong reciprocators would do better than those with low numbers under such circumstances. Strong reciprocators who punish defectors without regard to any future reward may significantly increase the survival chances of their group, under crisis conditions. The Gintis model is therefore essentially the same as Sober and Wilson's, where the greater fitness of strong reciprocator groups compensates for the within-group disadvantage these individuals suffer at the hands of selfish individuals. It also highlights the fact that moral behaviours, such as the punishment of anti-social behaviour, are just as likely to be the product of evolution as are behaviours that we would deem immoral or at least unethical. As Gintis's model shows, the balance between these two forces means that, at equilibrium, both selfish individuals and strong reciprocators co-exist – which is exactly what we see in Fehr's laboratory experiments.

Gintis suggests that the internalization of norms is the means by which strong reciprocity becomes established as a component of human behaviour. Internalization of norms refers to the way in which the older generation instils the values and rules of a culture into the younger generation. We have already discussed this process in Chapter 4, where we showed how, through the mind-shaping behaviours of adults, children come to appreciate not only that other individuals are mental agents with thoughts and beliefs of their own but also that a certain set of commonly held beliefs – Searle's institutional facts – is used to construct a social reality that wouldn't otherwise exist. This move to collective intentionality is achieved by the internalization of norms both by vertical transmission from parents to offspring and by oblique transmission from socializing institutions, such as schools and religions.

Gintis has shown, using mathematical models, that if the internalization of some norms has a fitness enhancing effect (for example, having good personal hygiene or the possession of a good work ethic) then genes promoting the capacity to internalize can also evolve. As niche construction theory argues, a culturally learned behaviour can then feed back on itself and have an effect on genes by acting as a source of selection. Gintis shows that, once a capacity to internalize has been established by a process of gene-culture co-evolution, then altruistic norms can also be internalized (as long as their fitness costs are not excessive). That is, even though strong reciprocity lowers the fitness of individuals relative to those who are selfish, such an altruistic norm can be internalized because it 'hitch-hikes' on the general fitness-enhancing capacity of norm internalization. Then, because groups with high numbers of strong reciprocators do better than those with low numbers, this offsets the fitness costs of adopting an altruistic norm.

social embeddedness

One of the outcomes of these kinds of studies has been to highlight the importance of social context. Humans decisions are embedded in social webs – networks of relationships that impose obligations. These relationships can have deep histories, in some cases reaching back several generations. The importance of social institutional effects of this kind also emerges very clearly from a large-scale study in which the same series of experiments was carried out in fifteen traditional societies around the world.

The 'Ultimatum Game' is one of the most widely used tests in experimental economics. In this game, one player is given a sum of money and has to make an offer to share it with a second (usually anonymous) player: the second player can accept the offer (in which case, the two players split the money according to the first player's offer) or refuse it (in which case, neither player gets anything). When this experiment is carried out in modern Western societies, offers typically average 50 per cent of the initial stake, suggesting that the player making the offer is responding to expectations of fairness (even though, in purely economic terms, he or she should make the lowest offer they think they can get away with). However, in the traditional societies, average offers ranged between 26–58 per cent. For this sample, two variables explained nearly 70 per cent of the variance in these data: first, the extent to which the group's economic production required co-operation (ranging from societies in which co-operation was limited to the immediate family's horticultural activities to ones where the mainstay of the economy was small boat whaling, which required large crews) and second, the extent to which the society concerned was integrated into (and thus dependent on) a market economy. The size of the offer made was positively (and independently) correlated with both variables.

The importance of social institutional factors emerges particularly clearly in the patterns of rejections. In almost all the experiments, offers that are considered too low are rejected by the second player. Typically, in studies of Western subjects, only offers below about 30 per cent of the stake are rejected, suggesting that players often seem to take the view that, even if there are limits, something is better than nothing. The sample of traditional societies showed much greater variability: in some cases, only offers below 16 per cent were rejected, whilst in others only offers above 70 per cent were accepted. This seemed to reflect social style. In societies where gift-giving was a matter of social honour and standing in the community, offers were rejected if they were too small (and hence considered an insult to the recipient); in other societies, accepting gifts (especially when unsolicited) establishes a relationship of obligation and future reciprocation and therefore even large gifts may be declined, if individuals are unwilling to place themselves in a position of obligation to a particular benefactor.

The social embeddedness of human interactions has led some authors to argue that the results reported in these economic experiments are simply a consequence of the fact that the human mind evolved to cope with life in small enclosed groups. We behave co-operatively towards strangers, because our minds are not adapted to life in modern, large-scale, societies. This view is implausible for several reasons, not least the fact that humans (and other apes and monkeys) readily differentiate between acquaintances and strangers. Indeed, female baboons not only recognise kin (as opposed to non-kin), but they also differentiate among kin according to the services they have to offer, varying their degree of co-operation through time. If female baboons are capable of such a finely-tuned response, it seems reasonable to expect humans to be capable of similar responses and not be limited to a crude rule of thumb.

More importantly, it is likely that human social networks have always been more extended than the initial impression sometimes given by our perception of life in small-scale societies. This point is made in Daniel Nettle's analyses of the size of language communities, which we discussed in Chapter 9: when ecological stability demands a wider network of co-operation, language communities (one marker for trading partnerships) are larger. A more explicit example is offered by Ernst Fehr and Joseph Henrich who point to the tradition of *hxaro* exchange relationships practised by !Kung San hunter-gatherers in southern Africa.

Hxaro relationships are long-term trading partnerships which help to manage environmental risk. During times of crisis, like droughts, individuals may 'activate' a partnership by travelling, as much as 200km, to visit one of their trading partners, staying there for several weeks and sharing water and food with the other members of the partner's group. Since different individuals sustained different *hxaro* partners, the various comings and goings would inevitably mean that strangers were often likely to be encountered in their *hxaro* partners' camps. (Reciprocal *hxaro* relationships did not extend to the other members of one's exchange partner's group, and there were no ties to the *hxaro* partners of these other group members.) Fehr and Henrich calculated that a single !Kung couple, with an average of 48 *hxaro* relationships between them, would range over an area of 10,000 km², potentially meeting up with more than one thousand people – a figure that, as we saw in Chapter 7, agrees well with the typical size of language groups (or tribes) in hunter-gatherer societies (and, indeed, *hxaro* relationships are always with members of the same tribe). The chance of a one-shot interaction with someone that they would never meet again would clearly be substantial, under such conditions. More importantly, perhaps, most of these individuals would not be known to one on a personal level – there would be no personal sense of obligation or trust towards them. On the other hand, it should also be remembered that *hxaro* partners would have been very concerned about their reputations under these circumstances: being rude or unhelpful to someone else's *hxaro* partner would be sure to land you in hot water with both the tribe member to whom the stranger was linked, as well as with one's own *hxaro* partner, since it would damage their relations with their own kinsmen. Even in these one-shot interactions, then, individuals would have remained embedded in a network of social obligation and on-going interactions.

These constraints aside, it is clear that the conditions under which *hxaro* partnerships operated are precisely those which Sober and Wilson specify for the operation of trait group selection: several non-isolated groups, individuals that disperse periodically so that groups change composition over time and groups that are in competition for resources. As we discussed in Chapter 2, the outcomes from such a process, especially when combined with cultural transmission of niche-constructing traits, can be markedly different from those of standard individual one-way models of evolution. Strong reciprocity is exactly the kind of trait that is predicted by this process.

Given the fission-fusion nature of human society, we can expect a number of other effects to be common. As Fehr and Henrich point out, the fact that one-shot interactions would have been (and, indeed, still are) common is a potent source of selection on individuals to learn to distinguish those who are friendly (and with whom one might wish to trade) from those who are hostile (who might have raiding or worse on their minds). Having cues that alert us to the kind of behaviour we can expect in another individual may thus be very important. The badges of group membership, that we discussed earlier, are examples of static, easy-to-read cues that provide us with a first estimate of another individual's reliability.

This may help to explain another feature of human behaviour that has been of considerable interest to social psychologists for many decades: the in-group/out-group effect. In modern large-scale societies, this is often reflected in the forms of behaviour that we identify as racist – discriminating against individuals on the basis of their physical appearance. A similar phenomenon exists in almost all traditional hunter-gatherer societies but in this case the distinction is usually drawn much more finely – between members of one's own tribe and everyone else. In many, if not most, such societies, the word which refers to members of the tribe is usually best translated as 'men' or 'humans'; those who belong to all other tribes are, by extension, considered to be 'not-men'.

It may be as well to remind ourselves once more of the conclusion we arrived at in Chapter 4: many aspects of human behaviour appear to be instinctive but this does not necessarily mean that they are hard-wired. Rather, they may appear to be instinctive because they are socially learned, very early in life. But the fact that they are learned does not, of itself, mean that they are easy to unlearn: as the Jesuits realized, things learned very early in life can become highly entrenched and resistant to change. Our moral sense is well developed, fully entrenched and operates even under the most artificial of laboratory settings. If there were one phrase that would best encapsulate what it means to be human it would perhaps be that we are a moral animal.

summary

Humans have a strong sense of morality and fairness. However, this highly characteristic feature of our species is something that evolutionary psychologists have only just begun to probe. In part this is because, in the main, biologists and psychologists believe that biological facts should not be used to derive moral sentiments (a view commonly known as the naturalistic fallacy). Strictly speaking, this view is based on a misunderstanding: whilst it is true that biological facts shouldn't solely determine whether or not something is moral, we do need these facts if we are to base our ethics and morals on a firm foundation. Studies of human economic behaviour in both Western and traditional societies reveal that people often engage in strong reciprocity, punishing cheats and paying a cost to do so, as well as fully expecting punishment if they cheat themselves. These kinds of moral behaviours are context-specific and can be tempered by the relative costs and benefits of punishment and the likelihood of on-going interaction. However, the fission-fusion nature of human society and the likelihood of one-shot interactions provided the conditions necessary for moral behaviours to evolve by a process of group-level selection and/or by a process of gene-culture evolution selecting for an ability to internalize norms.

glossary

Conformity bias the tendency to behave in the same or similar way to how other people are behaving in a given situation.

DNA Deoxyribonucleic Acid, the molecule that encodes genetic information and forms the basis of genetic inheritance

Dual inheritance theory a model of cultural evolution which views genes and cultural elements (memes) as separate forms of inherit-ance that need not necessarily interact with each other.

EEA (Environment of Evolutionary Adaptedness): the conglomeration of past environments, including environmental pressures, in which currently observed adaptations were shaped.

ESS (Evolutionarily Stable Strategy): a strategy that cannot be successfully invaded by any alternative strategy. The concept of an ESS recognizes that exactly what makes the best strategy depends on what everyone else in the population is doing. In this context, strategies may be either behavioural (a decision rule on how to behave, such as 'always punish those who defect on social contracts') or anatomical (such as the development of horns or other weapons).

Fertility the number of children produced by an individual in a given time period.

Fitness A measure of an individual's genetic contribution to future generations, relative to that of other individuals.

Free-rider an individual who takes advantage of the generosity of others by accepting the benefits of a social contract (or social living), but reneges on paying the associated costs.

Gossip hypothesis a theory for the evolution of language, which suggests that language evolved to bond large social groups. Building on the observation that monkeys and apes use grooming to bond their social groups, the gossip hypothesis claims that humans evolved language as a more efficient means of servicing social relationships in large groups.

Group selection a now discredited theory that evolution occurs for the 'good of the species (or group)'. Though widely held by biologists until the 1960s, group selection is in direct conflict with the principles of Darwinian evolutionary theory which assumes that selection occurs at the level of the individual (or, more strictly speaking, the gene).

Heritability the amount of phenotypic variation in a population that is the result of genetic differences between individuals in the population.

Imprinting a special case of learning, or 'programmed learning'; the process whereby a young animal becomes attached to another individual, usually its mother. Animals that have imprinted tend to attend and stay very close to, the animal they have imprinted on.

Intentionality a reflexively hierarchical scaling of belief states, defined by words such as *believe, suppose, imagine, assume, intend,* etc. First order intentionality is the capacity to have a belief about the contents of one's own mind; second order intentionality that of having a belief about someone else's mind state; and so on. Second order intentionality is equivalent to having **Theory of Mind** (q.v.).

Intersexual selection a form of sexual selection where female choice drives selection for male traits that are attractive to females

Intrasexual selection a form of sexual selection that is driven by same-sex competition for access to opposite-sex partners. Examples of intrasexually selected traits include large body size and weaponry, such as canine teeth.

Kin selection selection favouring altruistic acts between relatives, when the product of the benefit of the altruistic act to the recipient and the degree of relatedness is greater than the cost to the donor.

Lifetime Reproductive Success (LRS) the total number of living offspring that an individual contributes to the next generation.

Maladaptive a trait, character or behaviour which results in an organism possessing or performing it to have lower genetic fitness (q.v.) than one which does not. In the extreme case, it may result in premature death or the failure to reproduce.

Meme a term that refers to a unit of culture, analogous to gene.

Mind-reading the ability to understand the contents of another individual's mind (see Theory of Mind).

Natural selection Darwin's theory of the process by which evolutionary change occurs. Based on the principles of variation, inheritance and adaptation, the process of natural selection produces adaptations.

Naturalisic fallacy the principle that '*Is*' *does not mean* '*Ought*': a philosophical argument which states that we should not infer that a particular behaviour is 'good' or 'right' from the fact that it occurs, or is natural.

Neocortex the thin layer of neural tissue on the outside of the brain. The neocortex accounts for a large amount of brain volume in primates compared to other mammals, reaching as much as 80 per cent of total brain volume in humans.

Niche construction theory a theory which states that, rather than organisms being 'blindly' selected by the environment, in some cases organisms often modify the environments they occupy. In so doing, they set up a feedback loop for the action of natural selection.

Motherese an instinctive and distinctive style of speaking to young infants involving a higher pitched voice, a softening of intonation, large pitch contours and the use of short repetitive sentences.

Multi-level selection a theory for the evolution of altruism which states that selection operates not only on the individual but also at the level of the group in which the individual finds itself. Not to be confused with **group selection** (q.v.).

Mutualism behaviour that increases the fitness of both actor and recipient.

Parental investment any investment that parents make in an offspring which increases that offspring's chances of surviving. By definition, such investment imposes a cost to the parents as measured by their ability to invest in other offspring, current and future.

Phenotypic gambit the tactic used by research scientists to generate and test hypotheses about the adaptiveness of behaviour. The phenotypic gambit allows researchers to ignore the effects of other processes and so to focus on reproductive outcomes.

Prosocial attitudes or behaviours (such as generosity, forgiveness, etc.) which enhance the cohesion of social groups.

Prosody the melodic features of speech (for example, tone and pitch) that, combined with linguistic components, facilitate meaning and emotional content.

Reciprocal altruism a theory (based on the assumption that individuals take turns to exchange beneficial acts over a period of time) which explains how altruistic acts between unrelated individuals can evolve. In each exchange, the benefit of the acts to the recipient must be greater than the cost to the actor. Sometimes also known as *tit-for-tat*.

Reproductive value a measure of the average contribution to subsequent generations of individuals at any given age, relative to the contribution of the average individual.

Sexual selection a category of natural selection where the traits that are selected are those that increase an individual's likelihood of reproducing, rather than surviving. Sexual selection can operate **intersexually** or **intrasexually** (q.v.).

Social brain hypothesis the hypothesis which explains the evolution of large brain size, particularly in primates, as being driven by the need to solve complex social problems. It asserts that, rather than being driven by ecological problem solving, brain size evolved in response to the dynamic and sometimes unpredictable social world in which individuals are constantly forging and breaking alliances.

Theory of Mind (ToM) the ability to be aware of, and have a theory about, the thoughts, feelings, desires and intentions of other individuals. Theory of mind is believed to be a prerequisite for deception, imitation, and empathy. See also **Intentionality**.

bibliography

general sources

Badcock, C. (2000) *Evolutionary Psychology: a critical introduction*. Polity Press: London.

Barrett, L., Dunbar, R.I.M. & Lycett, J.E. (2002) *Human Evolutionary Psychology*. Palgrave-Macmillan: Basingstoke & Princeton University Press: Princeton.

Buss, D.M. (1999) *Evolutionary Psychology: the new science of the mind*. Allyn & Bacon: London.

Cartwright, J. (2000) *Evolution and Human Behaviour*. Macmillan: Basingstoke.

Cronk, L. (1999) *The Whole Complex: culture and the evolution of human behavior*. Westview Press: Boulder.

Dawkins, R. (1976) *The Selfish Gene*. Oxford University Press: Oxford.

Dunbar, R. (2004) *Grooming, Gossip and the Evolution of Language*. 2nd edition. Faber & Faber: London.

Dunbar, R. (2004) *The Human Story*. Faber & Faber: London.

Hrdy, S. (2000) *Mother Nature*. Oxford University Press: Oxford.

Plotkin, H. (1994) *Darwin Machines and the Nature of Knowledge*. Penguin: Harmondsworth.

Plotkin, H. (1997) *Evolution in Mind: an introduction to evolutionary psychology*. Penguin: Harmondsworth.

Plotkin, H. (2003) *The Imagined World Made Real: towards a natural science of culture*. Penguin: Harmondsworth.

Ridley, M. (1993) *The Red Queen: sex and the evolution of human nature*. Viking: London.

Wright, R. (1996) *The Moral Animal: why we are the way we are.* Abacus Books: New York.

chapter 1: why do we need evolution?

Darwin, C. (1859/1996) *On the Origin of Species.* Oxford University Press: Oxford.

Darwin, C. (1872/1998) *The Expression of the Emotions in Man and Animals.* HarperCollins: London.

Laland, K.N., & Brown, G. (2002) *Sense and Nonsense.* Oxford University Press: Oxford.

chapter 2: what evolution did for us

Foley, R.A. (1987) *Another Unique Species.* Longman: New York.

Klein, R. G. (1999) *The Human Career.* 2nd edition. Chicago University Press: Chicago.

Klein, R.G. & Edgar, B. (2002) *The Dawn of Human Culture.* Wiley: New York.

Laland, K.N., Odling-Smee, J. & Feldman, M. (2000) Niche construction, biological evolution and cultural change. *Behavioral and Brain Sciences* 23: 131–175.

Malik, K. (2000) *Man, Beast and Zombie: what science can and cannot tell us about human nature.* Weidenfeld & Nicolson: London.

Mameli, M. (2001) Mindreading, mindshaping and evolution. *Biology and Philosophy* 16: 597–628.

Plotkin, H.C. & Odling-Smee, J. (1981) A multi-level model of evolution and its implications for sociobiology. *Behavioral and Brain Sciences* 4: 225–268.

Sober, E. & Wilson, D.S. (1998) *Unto Others: evolution and psychology of unselfish behavior.* Harvard University Press: Cambridge.

Sterelny, K. (2003) *Thought in a Hostile World: the evolution of human cognition.* Blackwells: Oxford.

chapter 3: genes, development and instinct

Beauchamp, G.K., Katahira, K., Yamazaki, K., Mennella, J.A., Bard, J. & Boyse, E.A. (1995) Evidence suggesting that the odour types of pregnant women are a compound of maternal and fetal odour types. *Proceedings of the National Academy of Sciences, USA*, 92: 2617–2621.

Eliot, L. (1999) *What's Going on in There? how the brain and mind develop in the first five years of life.* Allen Lane: London.

Fernald, A. (1992) Human maternal vocalisations to infants as biologically relevant signals: an evolutionary perspective. In: J.H. Barkow, L. Cosmides, J. Tooby (eds) *The Adapted Mind,* pp. 391–428. Oxford University Press: Oxford.

Gottleib, G. (1971) *Development of Species Identification in Birds.* University of Chicago Press: Chicago.

Hepper, P.G. (1988) Foetal 'soap' addiction. *The Lancet* (June 11), pp. 1347–1348.

Kaitz, M. et al. (1987) Mothers' recognition of their new-borns by olfactory cues. *Developmental Psychology* 20: 587–591.

Monnot, M. (1999) The adaptive function of infant directed speech. *Human Nature* 10: 415–443.

Oyama, S. (1985) *The Ontogeny of Information.* MIT Press: Cambridge.

Ridley, M. (2003) *Nature via Nurture: Genes, experience and what makes us human.* Fourth Estate: London.

chapter 4: how to make us human

Barton, R.A. (1998) Visual specialization and brain evolution in primates. *Proceedings of the Royal Society, London,* 265B: 1933–1937.

Astington, J.W. (1993) *The Child's Discovery of the Mind.* Cambridge University Press: Cambridge.

Baron-Cohen, S. (1995) *Mindblindness: an essay on autism and Theory of Mind.* MIT Press: Cambridge.

Mitchell, P. (1997) *Introduction to Theory of Mind.* Arnold: London.

Perrett, D.I. and Emery, N.J. (1994) Understanding the intentions of others from visual signals: neurophysiological evidence. *Current Psychology of Cognition* 13: 683–694.

Tomasello, M. (2001) *The Cultural Origins of Human Cognition.* Harvard University Press: Cambridge.

chapter 5: choosing mates

Barber, N. (1995) The evolutionary psychology of physical attractiveness: sexual selection and human morphology. *Ethology and Sociobiology* 16: 395–424.

Cashdan, E. (1996) Women's mating strategies. *Evolutionary Anthropology* 5: 134–143.

Cornwell, R.E., Boothroyd, L., Burt, M.B., Feinberg, D.R., Jones, B.C., Little, A.C., Pitman, R., Whiten, S. & Perrett, D.I. (2004) Concordant preferences for opposite-sex signals? Human pheromones and facial characteristics. *Proceedings of the Royal Society, London,* 271B: 635–640.

Gangestad, S.W. & Thornhill, R. (1994) Facial attractiveness, developmental stability and fluctuating asymmetry. *Ethology and Sociobiology* 15: 73–85.

Grammer, K. (1989) Human courtship behaviour: biological basis and cognitive processing. In: A. Rasa, C. Vogel & E. Voland (eds) *The Sociobiology of Sexual and Reproductive Strategies,* pp. 147–169. Chapman & Hall: London.

Manning, J.T. (2002) *Digit Ratio: a pointer to fertility, behavior and health.* Rutgers University Press: Newark.

Milinski, M. & Wedekind, C. (2001) Evidence for MHC-correlated perfume preferences in humans. *Behavioural Ecology* 12: 140–149.

Nettle, D. (2002) Height and reproductive success in a cohort of British men. *Human Nature* 13: 473–491.

Nettle, D. (2002) Women's height, reproductive success and the evolution of sexual dimorphism in modern humans. *Proceedings of the Royal Society, London,* 269B: 1919–1923.

Pawlowski, B. (2003) Variable preferences for sexual dimorphism in height as a strategy for increasing the pool of potential partners in humans. *Proceedings of the Royal Society, London,* 270B: 709–712.

Pawlowski, B. & Dunbar, R.I.M. (1999) Impact of market value on human mate choice decisions. *Proceedings of the Royal Society, London,* 266B: 281–285.

Pawlowski, B. Dunbar, R.I.M. & Lipowicz, A. (2000) Tall men have more reproductive success. *Nature* 403: 156.

Penton-Voak, I. & Perrett, D.I. (2001) Male facial attractiveness: perceived personality and shifting female preferences for male traits across the menstrual cycle. *Advances in the Study of Behaviour* 30: 219–259.

Penton-Voak, I., Perrett, D.I., Castles, D.L., Kobayashi, T., Burt, D.M., Murray, L.K. & Minamisawa, R. (1999) Menstrual cycle alters face preferences. *Nature* 399: 741–742.

Singh, D. (1993) Adaptive significance of female physical attractiveness: role of waist-to-hip ratio. *Journal of Personality and Social Psychology* 65: 293–307.

Singh, D. (1994) Is thin really beautiful and good? Relationship between WHR and female attractiveness. *Personality and Individual Differences* 16: 123–132.

Voland, E. & Engel, C. (1990) Female choice in humans: a conditional mate selection strategy of the Krummhörn women (Germany, 1720–1874). *Ethology* 84: 144–154.

Wedekind, C. & Furi, S. (1997) Body odour preferences in men and women: do they aim for specific MHC combinations or simply heterozygosity? *Proceedings of the Royal Society, London,* 264B: 1471–1479.

chapter 6: the dilemmas of parenthood

Bereczkei, T. & Dunbar, R.I.M. (1997) Female-biased reproductive strategies in a Hungarian Gypsy population. *Proceedings of the Royal Society, London*, 264B: 17–22.

Boone, J.L. (1988) Parental investment, social subordination and population processes among the 15th and 16th Century Portuguese nobility. In: L. Betzig, M. Borgerhoff-Mulder and P. Turke (eds) *Human Reproductive Behaviour: a Darwinian perspective*, pp. 83–96. Cambridge University Press: Cambridge.

Cronk, L. (1989) Low socio-economic status and female-biased parental investment: the Mukogodo example. *American Anthropologist* 91: 414–429.

Crook, J.H. & Crook, S.J. (1988) Tibetan polyandry: problems of adaptation and fitness. In: Betzig et al. *Human Reproductive Behaviour*, pp. 97–114.

Daly, M. & Wilson, M. (1988) *Homicide*. Aldine de Gruyter: New York.

Dickemann, M. (1979) Female infanticide, reproductive strategies, and social stratification. A preliminary model. In: N.A. Chagnon & W. Irons (eds) *Evolutionary Biology and Human Social Behaviour: an anthropological perspective*, (pp. 321–368) Duxbury Press: North Scituate.

Emlen, S.J. (1995) An evolutionary theory of the family. *Proceedings of the National Academy of Sciences, USA*, 92: 8092–8099.

Kertzer, D.I. (1993) *Sacrificed for Honor: Italian infant abandonment and the politics of reproductive control*. Beacon Press: Boston.

Lycett, J.E. & Dunbar, R.I.M. (1999) Abortion rates reflect optimisation of parental investment strategies. *Proceedings of the Royal Society, London*, 267B: 31–35.

Salmon, C.A. & Daly, M. (1998) Birth order and familial sentiment: middle-borns are different. *Evolution and Human Behavior* 19: 299–312.

Sulloway, F. (1996) *Born to Rebel*. Pantheon: New York.

Trivers, R.L. & Willard, D. (1973) Natural selection of parental ability to vary the sex ratio. *Science* 79: 90–92.

chapter 7: the social whirl

Allman, J.M., Hakeem, A., Erwin, J.M., Nimchinsky, E. & Hof, P. (2001) The anterior cingulate cortex: the evolution of an interface between cognition and emotion. *Annals of the New York Academy of Sciences* 935: 107–117.

Barrett, L., Henzi, S.P. & Dunbar, R.I.M. (2003) From 'what now?' to 'what if?': *Trends in Cognitive Sciences* 7: 494–497.

Byrne, R. & Whiten, A. (eds) (1988) *Machiavellian Intelligence*. Oxford University Press: Oxford.

Cosmides, L. (1989) The logic of social exchange: has natural selection shaped how humans reason? Studies with the Wason Selection Task. *Cognition* 31: 187–276.

Deacon, T. (1997) *The Symbolic Species: The Coevolution of Language and the Human Brain*. Allen Lane: Harmondsworth.

Dunbar, R.I.M. (1993) The co-evolution of neocortical size, group size and language in humans. *Behavioral and Brain Sciences* 16: 681–735.

Dunbar, R.I.M. & Spoors, M. (1995) Social networks, support cliques, and kinship. *Human Nature* 6: 273–290.

Enquist, M. & Leimar, O. (1993) The evolution of co-operation in mobile organisms. *Animal Behaviour* 45: 747–757.

Hill, R.A. & Dunbar, R.I.M. (2003) Social network size in humans. *Human Nature* 14: 53–72.

Killworth, P.D., Bernard, H.P. & McCarty, C. (1984) Measuring patterns of acquaintanceship. *Current Anthropology* 25: 385–397.

Skoyles, J.R. and Sagan, D. (2002) *Up from Dragons: the evolution of human intelligence*. McGraw-Hill: Columbus.

Stiller, J. & Dunbar, R.I.M. (submitted). Perspective-taking and social network size in humans.

chapter 8: language and culture

Aunger, R. (ed) (2001) *Darwinizing Culture: The Status of memetics as a science*. Oxford University Press: Oxford.

Cavalli-Sforza, L., Feldman, M. W., Chen, K.H. & Dornbush, S.M. (1982) Theory and observation in cultural transmission. *Science* 218: 19–27.

Dunbar, R.I.M. (1993) The coevolution of neocortical size, group size and language in humans. *Behavioral Brain Sciences* 16: 681–735.

Dunbar, R.I.M. (2003) The social brain: mind, language and society in evolutionary perspective. *Annual Review of Anthropology* 32: 163–181.

Kinderman, P., Dunbar, R.I.M. & Bentall, R.P. (1998) Theory-of-Mind deficits and causal attributions. *British Journal of Psychology* 89: 191–204.

MacLarnon, A. and Hewitt, G. (1999) The evolution of human speech: the role of enhanced breathing control. *American Journal of Physical Anthropology* 109: 341–363.

Miller, G. (2000) *The Mating Mind*. London: Heinemann.

Mundinger, P.C. (1980) Animal culture and a general theory of cultural evolution. *Ethology and Sociobiology* 1: 183–223.

Searle, J. (1995) *The Construction of Social Reality*. Penguin, London.

Shennan, S. (2002) *Genes, Memes and Human History: Darwinian archaeology and cultural evolution*. Thames & Hudson: London.

Whiten, A., Goodall, J., McGrew, W.C., Nishida, T., Reynolds, V., Sugiyama, Y., Tutin,C.E.G., Wrangham, R.W. & Boesch, C. (1999) Cultures in chimpanzees. *Nature* 399: 682–685.

chapter 9: the uniqueness of human being

Boyd, R. & Richerson, P.J. (1985) *Culture and the Evolutionary Process.* Chicago University Press: Chicago.

Brown, P.J. (1986) Cultural and genetic adaptations to malaria: problems of comparison. *Human Ecology* 14: 311–332.

Cavalli-Sforza, L.L. & Feldman, M. (1981) *Cultural Transmission and Evolution: a quantitative approach.* Princeton University Press: Princeton.

Crook, J.H. & Crook, S.J. (1988) Tibetan polyandry: problems of adaptation and fitness. In: Betzig et al. *Human Reproductive Behaviour,* pp. 97–114.

Darley, J. & Latané, B. (1968) Group inhibition of bystander intervention in emergencies. *Journal of Personality and Social Psychology* 10: 215–221.

Durham, W.H. (1991) *Coevolution: genes, culture and human diversity.* Stanford University Press: Stanford.

Hinde, R.A. and Barden, L.A. (1985) The evolution of the teddy bear. *Animal Behaviour* 33: 1371–1373.

Henrich, J. (2001) Cultural transmission and the diffusion of innovations: adoption dynamics indicate that biased cultural transmission is the predominant force in behavioural change. *American Anthropologist* 103: 992–1013.

Henrich, J. & McElreath, R. (2003) The evolution of cultural evolution. *Evolutionary Anthropology* 12: 123–135.

Lumsden, C.J. & Wilson, E.O. (1981) *Genes, Mind and Culture.* Harvard University Press: Cambridge.

McGovern, T. H. (1981) The economics of extinction in Norse Greenland. In: T.M.L. Wrigley, M.J. Ingram & C. Framer (eds) *Climate and History,* pp. 404–433. Cambridge University Press: Cambridge.

chapter 10: virtual worlds

d'Aquili, E. & Newberg, A. (1999) *The Mystical Mind: probing the biology of religion.* Fortress Press: Minneapolis.

Atran, S. (2002) *In Gods We Trust.* Oxford University Press: Oxford.

Beit-Hallahmi, B. & Argyle, M. (1997) *The Psychology of Religious Behaviour, Belief and Experience.* Routledge: London.

Boyer, P. (2001) *Religion Explained: the human instincts that fashion gods, spirits and ancestors.* Weidenfeld & Nicholson: London.

Boysen, S.T., & Berntson, G.G. (1995) Responses to quantity: perceptual versus cognitive mechanisms in chimpanzees (*Pan troglodytes*). *Journal of Experimental Psychology: Animal Behavior and Processes* 21: 82–86.

Carroll, J. (1998) Literary study and evolutionary theory: a review essay. *Human Nature* 6: 119–134.

Dunbar, R.I.M. (2003) The social brain: mind, language and society in evolutionary perspective. *Annual Review of Anthropology* 32: 163–181.

Dunbar, R.I.M, Clark, A. & Hurst, N.L. (1995) Conflict and co-operation among the Vikings: contingent behavioural decisions. *Ethology & Sociobiology* 16: 233–246.

Hamilton, M. (2002) *Sociology of Religion.* 2nd edition. Routledge: London.

Hinde, R.A. (2000) *Why Gods Persist.* Routledge: London.

Jaynes, J. (1982) *Origin of Consciousness in the Breakdown of the Bicameral Mind.* Houghton Mifflin: New York.

Koenig, H.G. & Cohen, H.J. (eds) (2002) *The Link Between Religion and Health: psychoneuroimmunology and the faith factor.* Oxford University Press: Oxford.

Lewis-Williams, D. (2002) *The Mind in the Cave.* Thames & Hudson: London.

Newberg, A., d'Aquili, E. & Rause, V. (2001) *Why God Won't Go Away.* Ballantine Books: New York.

O'Connell, S. & Dunbar, R.I.M. (2003) A test for comprehension of false belief in chimpanzees. *Evolution and Cognition* 9: 131–139.

Rouget, G. (1985) *Music and Trance: a theory of the relations between music and possession.* University of Chicago Press: Chicago.

Sosis, R. & Alcorta, C. (2003) Signalling, solidarity, and the sacred: evolution of religious behaviour. *Evolutionary Anthropology* 12: 264–274.

Stiller, J., Nettle, D. & Dunbar, R.I.M. (2004) The small world of Shakespeare's plays. *Human Nature* 14: 397–408.

Thiessen, D. & Umezawa, Y. (1998) The sociobiology of everyday life: a new look at a very old novel. *Human Nature* 9: 293–320.

Whissell, C. (1996) Mate selection in popular women's fiction. *Human Nature* 7: 427–448.

Whitehouse, H. (2000) *Arguments and Icons: divergent modes of religiosity.* Oxford University Press: Oxford.

chapter 11: the science of morality

Barrett, L. & Henzi, S.P. (2002) Constraints of relationship formation among female primates. *Behaviour* 139: 263–289.

Barrett, L., Gaynor, D. & Henzi, S.P. (2002) A dynamic interaction between aggression and grooming among female chacma baboons. *Animal Behaviour* 63: 1047–1053.

Boyd, R. & Richerson, P. (1992) Punishment allows the evolution of co-operation (or anything else) in sizeable groups. *Ethology and Sociobiology* 113: 171–195.

Clutton-Brock, T.H. & Parker, G.A. (1995) Punishment in animal societies. *Nature* 373: 58–60.

Fehr, E. & Fishbacher, U. (2003) The nature of human altruism. *Nature* 425: 785–791.

Fehr, E. & Gächter, S. (2002) Altruistic punishment in humans. *Nature* 415: 137–140.

Fehr, E. & Henrich, J. (2003) Is strong reciprocity a maladaptation? In: P. Hammerstein (ed.) *The Genetic and Cultural Evolution of Cooperation.* MIT Press: Cambridge.

Fehr, E. & Rockenbach, B. (2003) Detrimental effects of sanctions on human altruism. *Nature* 422: 137–140.

Fehr, E., Fishbacher, U. & Gächter, S. (2002) Strong reciprocity, human cooperation and the enforcement of social norms. *Human Nature* 13: 1–25.

Gintis, H., Bowles, S., Boyd, R. & Fehr, E. (2003) Explaining altruistic behaviour in humans. *Evolution and Human Behavior* 24: 153–172.

Gintis, H. (2003) The hitchhikers guide to altruism: genes, culture and the internalization of norms. *Journal of Theoretical Biology* 220: 407–418.

Gintis, H. (in press) Solving the puzzle of prosociality. *Rationality and Society.*

Henzi, S.P., Barrett, L., Gaynor, D., Greef, J., Weingrill, T. & Hill, R.A. (2003) The effect of resource competition on the long-term allocation of grooming by female baboons: evaluating the priority of access model. *Animal Behaviour* 66: 931–938.

Milinski, M., Semman, D. & Krambeck, H.J. (2002) Reputation helps solve the 'tragedy of the commons'. *Nature* 415: 424–426.

Mealey, L., Daood, C. & Kruge, M. (1996) Enhanced memory for faces of cheaters. *Ethology and Sociobiology* 17: 120–127.

Nettle, D. (1999) *Linguistic Diversity.* Oxford University Press: Oxford.

Nettle, D. & Dunbar, R.I.M. (1997) Social markers and the evolution of reciprocal exchange. *Current Anthropology* 38: 93–98.

Orstrom, E., Gardner, R. & Walker, J. (1994) *Rules, Games and Common Pool Resources.* University of Michigan Press: Ann Arbor.

Sober, E. & Wilson, D.S. (1998) *Unto Others: evolution and psychology of unselfish behavior.* Harvard University Press: Cambridge.

Wilson, D.S., Dietrich, E. & Clark, A.B. (2003) On the inappropriate use of the naturalistic fallacy in evolutionary psychology. *Biology and Philosophy* 18: 669–682.

Wilson, D.S., Wilczynski, C., Wells, A. & Weiser, L. (2000) Gossip and other aspects of language as group-level adaptations. In: C. Heyes & L. Huber (eds) *The Evolution of Cognition*, pp. 347–365. MIT Press: Cambridge.

index